MAWTS-1

An Incubator for Military Transformation

ROBBIN LAIRD

EDWARD TIMPERLAKE

Copyright © 2024 by Robbin Laird and Edward Timperlake

All rights reserved.

No portion of this book may be reproduced in any form without written permission from the publisher or author, except as permitted by U.S. copyright law.

Library of Congress Control Number: 2024908293

Cover Photo: The first Marine Fighter Attack Squadron (VMFA) 314 "Black Knights" F-35C aircraft from Naval Air Station (NAS) Lemoore flown by CAPT Tommy Beau Locke from Strike Fighter Squadron (VFA) 125 "Rough Raiders" flies in formation over the Sierra's with the VFMA-314 squadron F/A-18A++, flown by LtCol Cedar Hinton aircraft "passing the lead" as part of the F/A-18 Sundown with the Black Knights. (U.S. Navy photo by Lt. Cmdr. Darin Russell/Released, June 5, 2019.

We would like to thank Chloe Ladd for editorial assistance in preparing this book for publication.

This book is dedicated to all of Ed's squadron mates who served together in VMFA-321 "Hells Angles".

The photo credit: Tom Duncavage, a noteworthy RIO and a brilliant aircraft combat paint scheme designer.

Contents

Foreword	ix
Preface	xiii
Prologue	1
Introduction	5

1. THE 2011 AND 2012 VISITS: THE COMING OF THE F-35B — 21
   - Preparing for the F-35B Transition: MAWTS Re-Shapes its Curriculum — 21
   - Another Aspect of the F-35 Transition: Rolling Out New Infrastructure — 25
   - The F-35 Squadron at Yuma: The Next Phase Begins — 26
   - Leveraging the F-35: MAWTS Prepares the Pilot of the Future — 31
   - General Amos on the F-35s at Yuma — 32
   - MAWTS and the Yuma F-35 Squadron: Evolving Capability into Operational Reality — 34

2. A 2013 PERSPECTIVE: THE YUMA INCUBATOR OF CHANGE — 38
   - The F-35B Comes to Yuma — 40
   - The Perspective of Lt. General (Retired) Trautman — 42
   - VMX-22 and the Yuma Incubator — 46

3. THE 2014 VISIT — 50
   - MAWTS-1 and Shaping the Future of USMC Aviation Within the Marine Corps — 51
   - Reflecting on the Tiltrotor Enabled Assault Force: The Perspective of a MAWTS-1 Osprey Instructor — 55
   - Visiting the F-35 Squadron at Yuma Air Station: The Executive Officer of VMFA-121 Provides an Update — 58

Working the Tactics and Training of F-35Bs with VMFA-121: The Perspective of Maj Roger "HASMAT" Greenwood	63
A VFMA-121 Maintainer Provides an Update on the Maintenance System for the F-35	66
4. THE 2016 VISIT	**70**
The Way Ahead for USMC Con-ops: The Perspective of Col Wellons, CO of MAWTS-1	72
Shaping a 21st Century Assault Force from the Sea: The Perspective from VMX-1	74
Working the MV-22 With F-35 Integration: Shaping Future TRAP Missions in a Dangerous World	77
The Green Knights On the Way to Japan: A Discussion with LtCol Bardo, CO, VMFA-121	80
5. THE 2018 VISIT	**85**
Col Wellons, MAWTS-1: Shaping a Way Ahead for the USMC and the Joint Force	85
MAWTS-1 Works Change: LtCol Ryan Schiller and His Team Discuss the Way Ahead	90
MAWTS-1 Works F-35 Integration: The Case of HIMARS	94
Integrating the ACE with the GCE in the Re-emergence of Great Power Competition	98
How to Prevail in a Disrupted and Degraded Combat Environment?	102
VMX-1 Working a Way Ahead for the MAGTF	104
Raising the Bar on Maintainability	107
6. 2020 INTERVIEWS AND VISIT	**113**
USMC Aviation Innovations for the 2020s	114
Colonel Gillette, CO of MAWTS-1, 2020	116
Working Mobile Basing	124
Moving Forward with Mobile Basing	126
The Role of Heavy Lift	131
Forward Arming and Refueling Points (FARPs)	134
MAWTS-1 and the F-35	137
The Evolving Amphibious Task Force	143
The Ground Combat Element in the Pacific Reset	147
Blue Water Expeditionary Operations and the Challenges for Aviation Ground Support Element	150

The F-35 and USMC-US Navy Integration	153
Unmanned Air Systems and the USMC	155

7. 2023 INTERVIEWS AND VISIT — 159

MAWTS-1 Works Mobile Basing and Support for the Distributed Joint Force	160
The Importance of Integration and Ownership in the Joint Fight: A Conversation with LtCol Barron at MAWTS-1	162
The Marine Corps Works the Next Phase of their Use of UAVs: The Perspective from MAWTS-1	164
Col Purcell's Perspective on the Impact of the Coming of the CH-53K	169
Operating from the Seafloor to the Heavens: MAWTS-1 Works the Spectrum Warfare Challenge	172
"Our excellence is a key part of deterrence"	175
ISR, Mission Planning and Enabling a Distributed Force: An Ongoing Challenge	177
The USMC Ground Combat Element and Shaping a Way Forward: Challenges to be Met	180
C3 and the Way Ahead for the USMC	184
Crafting a Sustainable Distributed Force: Maintenance and Logistics Challenges	186
Expanding the Assault Support Mission to a Broader Mission Set	190
Opening the Aperture on Force Integration	193
The Challenges of Working Expeditionary Advanced Base Operations	196
Refocusing the Force: MAWTS-1 Works on Ways Ahead	199

8. THE CHANGE OF COMMAND CEREMONY, MAY 2024 — 203

Looking Back at the Formation of MAWTS-1 and Shaping a Way Ahead	204
An Update on MAWTS-1: The Perspective of the Outgoing and Incoming Commanding Officers, May 2024	207

9. RETROSPECTIVES — 211

Recollections on the Establishment of MAWTS-1	211
Col Paul Boozman and the Creation of MAWTS-1	218

The Perspective of MajGen John Cox: Present at the Creation of MAWTS-1	222
Institutionalizing Excellence: The Raison d'être for Creating MAWTS-1	224
The Perspective of the First Commander of MAWTS 1: LtCol Howard DeCastro	226
The Second Commander of MAWTS-1: Bob Butcher	227
The Third Commander of MAWTS-1: Major General C.L. Vermilyea	230
The Fourth Commander of MAWTS-1: Randy Brinkley	233
How MAWTS-1 Works on Combat Innovation: A Look Back with Col Michael Kurth	237
MAWTS-1 and Trajectory Vision: Talking with the 5th Commander of MAWTS-1	244
The 8th Commander of MAWTS-1: LtGen Barry Knutson Jr.	246
The Perspective of the 9th Commander of MAWTS: LtGen Castellaw	249
The Perspective of the 13th Commander of MAWTS-1: MajGen Raymond Fox	251
Turning Training into New Combat Capabilities	254
What MAWTS-1 Contributes to the Force	256
A Conversation With the 14th Commander of MAWTS-1: LtGen "Dog" Davis	258
10. LTCOL DECASTRO SUMS IT UP	262
Appendix: The Commanders of MAWTS-1	265
About the Authors	279
Other Second Line of Defense Books of Interest	281
Notes	289

Foreword

In their new book, *MAWTS-1 An Incubator for Military Transformation*, Robbin Laird and Ed Timberlake have captured the essence of Marine Aviation Weapons and Tactics Squadron One (MAWTS-1) in a very creative way.

They begin with a wide-ranging series of poignant interviews with the squadron's recent Commanding Officers and highly skilled instructor pilots going back to introduction of the F-35 to MCAS Yuma in 2011.

These interviews are illuminating because they expose the role MAWTS-1 plays in keeping Marine aviation moving forward. New concepts of operation and new weapon systems demand refined tactics and innovative methods of training.

As the Marine Corps exploits the incredible capabilities of the V-22 Osprey, F-35 Lightning II, CH-53 King Stallion, the TPS-80 Ground/Air Task Oriented Radar (G/ATOR) and the Common Aviation Command and Control System (CAC2S), the MAWTS-1 staff, and the Weapons and Tactics Instructors (WTI's) they produce, are at the cutting edge of keeping the Fleet Marine Force ready and relevant.

Over the past four years, the task of introducing new aviation

Foreword

weapons systems has become exponentially more challenging as a direct result of the 38th Commandant's sweeping changes as delineated in Force Design 2030 (FD 2030).

This initiative, which began in 2019, cut a significant portion of the Corps' active-duty aviation force structure and slowed production procurement ramps significantly without any commensurate reduction in operational commitments.

Fewer squadrons, less aircraft, and the forced introduction of new operational concepts not ideally suited to the previously preferred Marine Air Ground Task Force (MAGTF) model placed a significant burden on Marine Corps aviation – but the Marines of MAWTS-1 proved they were up to the challenge by refining the tactics, techniques, and procedures needed to accommodate the guidance they had been given.

How did MAWTS-1 evolve to become such an indispensable pillar within the Marine Corps?

Its significance lies in its role as the premier training unit for Weapons and Tactics instruction, where it refines and disseminates cutting-edge concepts of employment – often with new weapons that have potential capabilities never previously imagined.

By continually adapting to emerging threats and technological advancements, MAWTS-1 ensures that Marine aviators are equipped with the skills and knowledge necessary to excel in the ever-evolving landscape of modern warfare.

The squadron's contributions extend beyond training, influencing the development of new doctrines and fostering a culture of innovation within the Marine Corps aviation community. In essence, MAWTS-1 serves as a linchpin in maintaining the Marine Corps' tactical edge, preparing its aviators for the challenges of today and the uncertainties of tomorrow.

After the authors track the recent evolution of the squadron, they double back with a series of interviews with the pioneers and Commanding Officers who developed and fostered the MAWTS-1 concept, fought off the naysayer's resistance to change, and led Marine aviation into what it has become today – an essential

Foreword

element of the MAGTF dedicated to bringing about a revolution in next generation aviation capabilities.

These retrospectives may be the most valuable part of the book because they showcase the impact a small cadre of individuals, disappointed in their assessment of post-Vietnam Marine aviation, can have. Their visionary ideas set a new course that grew from something innocuously called "Project 19" into the premier aviation training squadron in the world – MAWTS-1.

In the final chapter of the book, the authors include a speech given by the first Commanding Officer of MAWTS-1, Lieutenant Colonel Howard DeCastro, to the squadron at a Marine Corps Ball in 2019 in which he says:

After General Amos retired as Commandant, he and I had a chance to talk. He told me that MAWTS literally saved Marine Corps aviation. I suspect that was an exaggeration, but, without question, MAWTS has dramatically improved Marine aviation and our working relationship with the Ground Forces.

My personal experience dating back to my time as a WTI student in one of the first MAWTS-1 classes ever conducted, then as a member of the instructor staff, followed by tours in command at the squadron, group, and wing level, and ultimately as the USMC Deputy Commandant for Aviation is that Colonel DeCastro and the authors have it right – MAWTS-1 is indeed An Incubator for Military Transformation, and this book tells the squadron's story in a fascinating and compelling way.

George J. Trautman III

LtGen, USMC (Ret)
Former USMC Deputy Commandant for Aviation

Preface

Training for military forces is in the throes of significant change. The threats are dynamic; there is always the reactive enemy; and technology fosters new ways to operate.

Concepts of operations are evolving, most notably as U.S. and allied forces are focusing on force distribution to deal with the higher end threats authoritarian adversaries are fielding.

We (Robbin Laird and Ed Timperlake) have visited the major training centers in the United States and several abroad as the state of the art of training is dynamically developing as well.

In this book, we highlight our visits to a major training center, MAWTS-1 located at Marine Corps Air Station, Yuma. This is a truly multi-domain training center and has been from its inception.

We have visited together or separately several times since 2011 so have seen the introduction of new air systems into the USMC and have been able to talk with Marines as they shifted from preparing for the land wars to the higher end fight generated by authoritarian adversaries.

We have then had the benefit of talking with several former commanders of MAWTS-1 to gain further understanding of how MAWTS-1 was established and has evolved throughout its history.

Without their insights, the book would not have been complete and we thank them for their contribution, and not just to the book.

With regard to the interviews we have conducted from 2011 through 2024, we have indicated the date the interview was published on one of our websites, either *Second Line of Defense* or *Defense.info*. These interviews were conducted at the time and provide insights at the time of the interviews regarding how the Marines were working their training regime. It is not the case that we are simply remembering; these interviews were published at the time indicated.

As a final thought, we might look back at October 23 1983 when Marine forces ashore from Battalion Landing Team 1/8 in Beirut, Lebanon were attacked. The attackers employed a suicide driver of a fully loaded truck bomb to take down the building they had occupied.

Two hundred and twenty marines, eighteen sailors and three soldiers were killed in action by the surprise attack. A group called the Islamic Jihad claimed responsibility but it was widely thought that Iran through Hezbollah had culpability. It was the largest single mission death toll since the Battle of Iwo Jima.

The American military is brilliant in investigating the reasons why combat engagement go the way they do. In the case of a contributing factor leading to the enemy's success a tragic sentence stands out that on the security perimeter the Rules of Engagement directed safety over security because the units on guard duty were not allowed to have a round chambered in their weapons.

Such ROE stupidity cost precious time as the truck raced into the building with reported 12,000 pounds of TNT.

After the attack a U.S. Navy Carrier task force off the coast was given an alert order to stand-by to launch an air strike. On 3 December the carrier air wings of USS Independence and Kennedy were given an alert order to attack six specific targets in as reported a four hour launch.

They had previously been told to stand down from constant alert on December 1st.

The higher command over guidance, smacking of President

Preface

Johnson saying my boys could not bomb an outhouse unless I told them, rang down as a sad deadly legacy of ready fire aim, asking the aircrews to fly into heavy defensives without their seasoned combat insights having influence.

Attacking into the rising sun against some worthless targets with tremendous triple A and ground to air missiles was a disaster. An A-6 was shot down with a pilot killed and an A-7 flying a rescue cap was hit with pilot ejecting feet wet. The legacy of who picked the time on target, it was reported "higher command" even miscalculated the time zones and worthless target set is lost to history.

But that Navy air combat event was not lost on Secretary of the Navy John Lehman who had flying experience as a BN, Bombardier/ Navigator in the A-6 Intruder.

He ordered the creation of "Strike U" which officially is designated Naval Strike and Air Warfare Center. As reported by Col Randy "Dragon" Brinkley, MAWTS was a very strong building block in advancing Navy integrated Fleet training and tactics.

We visited NAS Fallon and wrote a series of reports and learned of a change of name from Navy Fighter Weapons School to Naval Aviation Warfighting Development Center in 2015.

Here was the slide presented in a discussion with us during a 2015 visit to NAWDC:

UNCLASSIFIED

 NAWDC Evolution

- 1969 – Navy Fighter Weapons School - TOPGUN
- 1984 – Naval Strike Warfare Center
- 1988 – Carrier Airborne Early Warning Weapons School
- 1996 – Naval Strike and Air Warfare Center (NSAWC)
- 1998 – Rotary Wing Weapons School - SEAWOLF
- 2000 – Electronic Reconnaissance Weapons School
- 2001 – Joint Close Air Support School
- 2011 – Growler Weapons School - HAVOC
- 2012 – Tomahawk Land Attack Missile Cell
- 2015 – Naval Aviation Warfighting Development Center

Born and Adapted Through Failures in Combat

Naval Aviation Warfighting Development Center   Fly, Fight, Lead – Win

Preface

This change of name is very significant and represents a culmination of the work of the Top Gun era and the command directed foundation for the integrated war fighting approach for the distributed fleet.

In one of our visits, we were briefed on a specific example of NAWDC as a USN incubator of change. Instructors were asked in real time to engage with an Air Wing on station tasked with night air-to-ground strafing attacks in the unforgiving mountains of Afghanistan. They were asked at NAWDC to develop in real time better tactics for such an operation.

That is a perfect example of todays Navy communicating with actionable intelligence and fighting at the speed of light half a world away.

Training is not simply learning what we have done: it is what we can and should do in current operations and anticipating the future fight. MAWTS-1 was established with this approach in mind and it continues as its core task.

But new technologies are being introduced in the training regime such as Live Virtual Constructive training, and much greater emphasis on the tools of the trade to fight a peer competitor such as signature management, counter-ISR, cyber war, and many other new techniques.

But war has not changed in a key essential: it is about force being applied to achieve a nation's objectives and to protect its sovereignty. In a period in which multi-polar authoritarianism is on the march, being able to defend our interests and way of life is crucial. The training at MAWTS-1 is a key part of enhancing our ability to do so.

We as citizens of Western democracies owe much to the men and women who come to MAWTS-1 and go from there to deploy and serve their country. We salute them and have written this book to honor them. We hope our readers will enjoy the journey.

We would especially like to thank LtGen Steve Rudder (Ret.) for suggesting the interviews with the previous commanders of MAWTS-1 and then going the extra mile to rounding them up which allowed us to organize these unique interviews.

Prologue

After returning to the United States from Australia on April 25, 2024, I buckled up again for a flight to Yuma, Arizona to have the privilege of witnessing the MAWTS-1 change of command ceremony.

In the process of finishing up our book on MAWTS-1, this seemed a good way to close out our effort.

One of the architects of the MAWTS-1/WTI concept and the first MAWTS-1 Commanding Officer, LtCol Howard DeCastro, had written to me suggesting the idea, and the CO of MAWTS-1, Col Purcell kindly agreed to invite me.

This would give me the chance to meet with several of the previous Commanding Officers of MAWTS-1, meet the three 3 Star USMC Generals attending the ceremony and meet the new CO of MAWTS-1 as well.

On the day before the ceremony, I had the chance to sit down and interview two of the first commanding officers, LtCol DeCastro and LtGen Barry Knutson. In the afternoon, I was able to interview the outgoing CO, Col Purcell, and the incoming CO Col Joshua Smith.

Col Smith on the left and Col Purcell on the right at the change of command ceremony, May 3, 2024.
Credit Photo: Robbin Laird

What was amazing about the two sets of interviews is how connected in time they were.

The first and eighth CO of MAWTS focused on the approach they built towards combat innovation, namely, inserting technology into con-ops rather than having technology existing outside of the organizational changes needed to use relevant technologies.

It was the warfighters driving innovation in terms of real warfighting improvements, rather than some contractor or acquisition official pushing technology down their throats.

Then two hours later, I had the same conversation with Purcell and Smith.

It was about technology that did not exist at the time when DeCastro and Knutson were in charge, but it was the same mentality and same drive for combat excellence which we discussed.

MAWTS-1

Col Purcell returns from his last flight as CO of MAWTS-1. Credit Photo: Robbin Laird

And I would conclude with just one thought – don't change the course.

The drive for warfighting excellence in the operating force is not nice to have, it is what we need if our country continues to field a warfighting force respected by the world, by both allies and adversaries.

Well I am not a Marine, but it is hard to not listen to the USMC hymn at the ceremony and not say Semper Fidelis. And working with Ed Timperlake, I have certainly learned there is no such thing as a former Marine.

By Robbin Laird

Introduction

The United States Marine Corps (USMC) Vietnam air ground war was the experience which shaped the founding of MAWTS-1 or the past was prologue for the formation of MAWTS-1.

The historical accounting for total aircraft and helicopter losses in combat in Southeast Asia during the Vietnam War has been listed in many after action historical surveys as close to 10,000. Fixed wing losses are slightly fewer than helicopters.

The aircraft with the highest causality rate was the F-4 Phantom II flown by Air Force, Navy and Marine aircrews. Over 5,000 were produced worldwide and approximately 700 lost in Southeast Asia with over 500 of those being direct combat losses.

For our aircrew lost in combat, and also sadly many tragic accidents because of demanding operational tempo in unforgiving weather and terrain, theirs was a lasting story of the ultimate sacrifice of courageous aircrews. Those aviators showed undaunted courage.

For combat aviators in all services Navy, Air Force and Marines, enough was enough.

Airpower leaders who survived and stayed in uniform after Vietnam made it their quest to identify significant deficiencies in training, tactics, and combat employment of airpower during the Vietnam War and to change the way training was done. They wanted to focus on training for reality and how to prevail, not simply to train.

An F-4 Phantom II flown by a fighter crew serving in VMFA-321 "Hells Angels" launch an AIM-7 Sparrow missile " This was Ed Timperlake's Squadron's Christmas Card for one year. Photo credit: Art "Alpha" Brooks USMC RIO serving in VMFA-321.

The Vietnam experience was key to the spirit which underlay the need to develop the new warfighting development centers. That experience was rooted in the gap between the reality on the ground and how the war was conducted,

Suggestive of this gap was a brilliant spoof which was developed in 1966 by USAF fighter pilots flying the Phantom II with the 12th Tactical Fighter Wing out of Cam Ranh Bay Vietnam.

They set up a mock interview between an F-4 pilot and a reporter with the ubiquitous command political "spin" information officer sitting in to correct the pilots blunt truthful language. It was called "What the Captain Means" and the ending quote was insightfully prescient for what was to follow for almost a decade long war.

Reporter: "I noticed in touring the base that you have aluminum matting on the taxiways. Would you care to comment on the effectiveness and usefulness in Vietnam?"

Pilot: "You're friggin right I'd like to make a comment. Most of us pilots are well hung, but ****, you don't know what hung is until you get hung up on one of those friggin bumps on that goddamned stuff."

Public Affairs Officer: "What the Captain means is that the aluminum matting is quite satisfactory as a temporary expedient, but requires some finesse in taxying and braking the aircraft."

Reporter: "Did you have an opportunity to meet your wife on leave in Honolulu, and did you enjoy the visit with her?"

Pilot: "Yeah, I met my wife in Honolulu, but I forgot to check the calendar, and so the whole five days were friggin well combat-proof. A completely dry run."

PAO: "What the captain means is that it was wonderful to get together with his wife and learn first-hand about the family and how things were at home."

Reporter: "Thank you for your time, Captain."

Pilot: "Screw you, why don't you bastards print the real story instead of all that crap."

PAO: "What the Captain really means is that he enjoyed the opportunity to discuss his tour with you."

Reporter: "One final question. Could you reduce your impression of the war into a simple phrase or statement, Captain?"

Pilot: "You bet your ass I can. It's a f**ked-up war."

PAO: "What the Captain means is it's a f**ked-up war."[1]

With such a funny but searing indictment of how the air ground war was fought each service with airpower assets came up with their solutions to incubate combat excellence and to shape a new direction of combat excellence.

The U.S. Navy, including Marine aircrews, came up with a Navy air-to-air fighter weapons school immortalized as "Top Gun". In a brief history of Top Gun published by the Department of Defense, the reason for establishing the new Navy warfighting center was highlighted:

During the Vietnam War, Navy fighter pilots and aircrew were dying at an alarming rate," explained Navy Cmdr. Dustin Peverill, a 20-year Navy veteran and two-time TOPGUN instructor. "The Navy was losing a lot of airplanes and, more importantly, a lot of aircrew."

Despite having the technological edge, the Navy was experiencing unacceptable combat losses in Vietnam. In response, the service commissioned an investigation and tasked Navy Capt. Frank Ault to lead the effort.

The resulting report, known as the Ault Report, highlighted many performance deficiencies and their root causes, including the need for an advanced course to teach fighter tactics. The result was the Navy Fighter Weapons School, established at Miramar in 1969.

Nicknamed TOPGUN, the school's mission was — and still is — to train aircrew in all aspects of aerial combat to be carried out with the utmost professionalism. In its early days, its students were trained over the course of four weeks on F-4 Phantom II aircraft to get better at one-on-one aerial combat, also known as dogfighting.

"When TOPGUN graduates began to go back to the fleet in the early 1970s and the air war started back up, the Navy's kill ratio jumped. TOPGUN worked," Peverill said. "It validated that the training, the subject-matter expertise and, most importantly, the professionalism that it produced worked in combat and it produced results."[2]

The USAF created a brilliant multi-aircraft extremely demanding with safety stressed exercise that is known as "Red Flag."

According to the 414th Combat Training Squadron:

RED FLAG was established in 1975 as the brainchild of Lt. Col. Richard "Moody" Suter and one of the initiatives directed by General Robert J. Dixon, then commander of Tactical Air Command, to better prepare our forces for combat.

Lessons from Vietnam showed that if a pilot survived his first 10 combat missions, his probability of survival for remaining missions increased substantially. Red Flag was designed to expose each "Blue" force pilot to their first 10 "combat missions" here at Nellis, allowing them to be more confident and effective in actual combat.

This same principle continues to guide Red Flag today, with the goal of preparing Air Force, Joint, and Coalition pilots, aircrew and operators to fight against a near-peer adversary in any combat environment.

Tasked to plan and control this training, the 414th Combat Training Squadron's mission is to maximize the combat readiness, capability and survivability of participating units by providing realistic, multi-domain training in a combined air, ground, space and electronic threat environment while providing opportunity for a free exchange of ideas between forces.

MAWTS-1

Aircraft and personnel deploy to Nellis for RED FLAG under the Air Expeditionary Force concept and make up the exercise's "Blue" forces.

By working together, these Blue forces are able to utilize their diverse capabilities and weapons systems to execute specific missions, such as offensive counter air, suppression of enemy air defense, combat search and rescue, dynamic targeting, and defensive counter air.

These forces use various tactics to attack NTTR targets such as mock airfields, vehicle convoys, tanks, parked aircraft, bunkered defensive positions, missile sites, and conduct personnel recovery efforts. These targets are defended by a variety of simulated "Red" force ground and air threats to give participant aircrews the most realistic combat training possible.

The Red force threats are aligned under the 57th Operations Group, which controls seven squadrons of USAF Aggressors, including fighter, space, information operations and air defense units.

The Aggressors are specially trained to replicate the tactics and techniques of potential adversaries and provide a scalable threat presentation to Blue forces which aids in achieving the desired learning outcomes for each mission.

A typical RED FLAG exercise involves a variety of attack, fighter and bomber aircraft (F-15E, F-35, F-16, F/A-18, A-10C, B-1B, B-2A, B-52H, FGR4, MQ-9, etc.), reconnaissance aircraft (MQ-4B, RC-135, U-2S), electronic warfare aircraft (EC-130H, EA-18G and F-16CM), air superiority aircraft (F-22A, F-15C, etc), airlift support (C-130, C-17A), Search and Rescue aircraft (HH-60G, HH-60W, HC-130J, CH-47), aerial refueling aircraft (A330, KC-130, KC-135R, KC-10A, KC-46A, etc), multi-domain Command and Control platforms (E-3, E-8C, E-2C, E-7A, R1, etc) as well as ground based Command and Control, Space, and Cyber Forces.

Four U.S. military services, their Guard/Reserve components and the air forces of numerous other countries participate in each RED FLAG exercise. Since 1975, 29 countries which includes (EPAF a consortium of Netherlands, Belgium, Denmark and Norway) and NATO (AWACS) have joined the U.S. in these exercises and several other countries have participated as observers.

RED FLAG has seen 30,268 aircraft and has provided training for more than 529,722 military personnel, of which 164,724 are aircrew members flying more than 423,248 sorties and logging more than 783,907 hours of flying time.[3]

Robbin and I have visited both training centers several times in the past and have seen how they are evolving training now for today's high-end fight.

Shaping a Way Ahead for Combat Excellence

After Vietnam, the Marine Corps created their own weapons training squadron: MAWTS-1, the Marine Air Weapons Training Squadron based at MCAS Yuma Az. In this book, we tell their story.

The key driver for combat learning is to trust those who put it all on the line and empower them to band together to share and never stop learning. It is often said in the air, 'ability leads not rank'.

The key dynamic for having success in airpower engagements in both air-to-air and air-to-ground missions begins with trusting the insights of squadron fighter pilots as they progress through the ranks. The very good ones, the best of the best, always want to leave a legacy for those that follow.

The first commanding officer at MAWTS-1 was LtCol Howard DeCastro. His combat awards include a Legion of Merit (non-combat), a Distinguished Flying Cross (DFC) and 20 Air Medals. He was selected as the "Marine Aviator of the Year," receiving the Alfred A. Cunningham Award named for the first Marine aviator.

We have interviewed him and several of the previous commanders of MAWTS-1, and their interviews conclude this book. Our concluding piece is the Marine Birthday Ball speech DeCastro gave in Yuma in 2019.

As he underscored in that speech:

The challenge for us 41 years ago, and the challenge for you today, is to continually improve. To improve takes an understanding of enemy capabilities and projecting their future capabilities so we can develop the hardware, software, tactics, and our own capabilities that keep us always in the lead.

Every one of you should be proud to have earned the right to be assigned to MAWTS, and you should be proud of what you are accomplishing.

I was going to give you a motivational speech, but then I realized you are just like the original members of MAWTS-1.

You are completely self-motivated, are never satisfied with the status quo, and you will keep thinking, keep communicating, keep innovating, and keep pushing each other to make Marine Air and the Marine Air Ground Team better every day. That is who you are, and you know how important it is to keep getting better.

You know that the Marines on the ground trust you and are counting on you. For everything you have done and all you do, I thank you!

The core point really is this: *You are completely self-motivated, are never satisfied with the status quo, and you will keep thinking, keep communicating, keep innovating, and keep pushing each other to make Marine Air and the Marine Air Ground Team better every day.*

That is what MAWTS-1 success in training is based on, namely the drive to excellence and the innovation of the squadron pilots. The idea from the outset was not to prioritize fix wing pilots to lead the command but to broaden the leadership scope to include rotary wing and later tiltrotor commanders as well.

"Alpha" and "Easy" Ed Timperlake go bombing" and open canopies.

I wrote a piece in 2016 which I entitled: *Squadron Fighter Pilots: The Unstoppable Force of Innovation for 5th Generation Enabled Concepts of Operations*. This is the spirit and driving force of innovation at MAWTS-1 – the squadron pilots across the flying force. In that article I highlight:

The skillfulness and success of fighter pilots in aerial combat is an extensively researched yet modestly understood and fundamentally complex concept.

Innumerable physical and psychological factors along with chance opportunities affect a pilot's facility for success in air combat.

Perhaps the best narrative of the intangibles of the skill and courage of a fighter pilot was captured by the author Tom Wolfe in his seminal work The Right Stuff.

"From the first day a perspective fighter pilot begins their personal journey to become a valued and respected member of an elite community, serving as an operational squadron pilot, the physical danger is real.

"But so is the most significant force for being the absolute best that a fighter pilot can feel which is day in and day out peer pressure by those they really and truly respect, their squadron mates."[4]

Put it in the hands of the warfighters and let them drive the innovations needed for the fighting force is a key to the kind of training which goes on at MAWTS-1. As a former Commanding Officer of MAWTS-1 and a former Deputy Commandant of Innovation, LtGen "Dog" Davis put it:

"I think it's going to be the new generation, the newbies that are in the training command right now who are getting ready to go fly the F-35, who are going to unleash the capabilities of this jet. They will say, 'Hey, this is what this system will really give me. Don't cap me; don't box me in."[5]

Senior combat pilot commanders, many coming out of the Vietnam war, albeit much junior at the time, achieved two magnificent victories by both winning the air rivalry against the USSR in the Cold War and achieving historic air combat success in the magnificent air campaign of Desert Storm.

The lesson for the air power rivalry between the U.S. and USSR is rather straightforward: the technology had to be available, but it also had to be successfully understood and employed.

A historical take away from the cold/hot war air battles is that in the air-to-air mission a country that equips its fighters with airborne radar and sensors allows more autonomous action and favors tactical simplicity and operational autonomy—even though the equipment becomes more complex.

In air-to-ground, airborne simplicity indicators are usually smaller formations and allowance to maneuver independently into weapon launch envelopes primarily in a weapons-free environment. Embedding technology into the weapon itself –bombs and rocket-fired weapons– has also made a revolutionary difference.

In air combat a nation must always assume a reactive enemy can develop the necessary technology to try and mitigate any advantages. With the worldwide proliferation of weapons even a second or third world nation might have state-of-the art systems. The air war in Vietnam was a technology peer-to-peer war.

In the book *Fighter Pilot* by his daughter Christina Olds, the story is told of Gen. Robin Olds, a Triple Ace with 16 kills in aerial combat that began in WWII chasing down Nazi Luftwaffe and Messerschmitt. Olds was next appointed to Command the famous 8th TAC Fighter Wing out of Udorn and Ubon, Thailand. These are the "Top Guns" of the Air Force. The 8th TAC Fighter Wing called themselves the "Red River Rats," while the Navy and Marine pilots often wore patches calling themselves "Yankee Air Pirates", it was that kind of war.

Olds flew into "the 9 gates of Hell," aka North Vietnam, no less than 101 times dog-fighting enemy fighter jets, including Russia's notoriously fierce MiG 21. Once crossed into North Vietnam, these top Air Force, Navy, and Marine fighter crews faced staggering odds of making it back every time, but they flew into North Vietnam repeatedly anyway.

As Olds reported, *There were more anti-aircraft guns within a 60-mile radius of Hanoi than Germany had possessed in all of Europe.*

The air war over the skies of Vietnam and in the Middle East in the Yom Kippur War was between two aviation technology peer competitors because of USSR TacAir type/model/series (T/M/S) support to aerial advisories.

Stephen Ambrose in his award-winning book *Citizen Soldier* about the U.S. Army fighting in Europe during WWII made a brilliant point: intelligence does not make decisions. Decisions are made by the mindset of those receiving the intelligence.

Consequently, the lesson on the Cold War U.S.-USSR rivalry is that air combat leaders must be able to adjust during the course of an air battle or war by changing strategy and tactics, to achieve exploitation of the enemy's mistakes or weakness.

Aircrews must be adaptable enough to follow changing commands from leadership and, on their own initiative, change tactics to achieve local surprise and exploitation. Like the quote in Animal House, "knowledge is good," in the cockpit, it can be a lifesaver and aid in a mission accomplished.

Among the most intangible qualities of a combat force are those cultural factors that influence its basic fighting capabilities. These qualities can be of paramount importance.

To take what is the most sensational example, consider the Kamikaze pilot. No mere quantitate assessment of the Japanese tactical aviation forces of the Second World War could have accounted for Kamikazes. Only an assessment of cultural characteristics could have keyed analysts to the possibility.

In retrospect, we can understand that the Japanese belief in the divinity of their empire and the cultural abhorrence of shame could allow for creating pilots sufficiently motivated to embrace suicidal missions.

The example of Kamikazes is not representative of this discussion, but only illustrates that cultural factors, despite their intangibility, must somehow be reckoned with.

One of the essential elements of creating a successful combat air crew is simply motivation often expressed as dedication, heart, will, ambition or competitiveness. It captures the qualities of a fighting force that makes its warriors enthusiastic rather than lackadaisical or dispirited.

Of course, inside the ever-advancing complexities of 21st Century 5th Gen aircraft technology and the resulting con-ops, fighter pilots must have the capacity to understand and operate the sophisticated technology of their state-of-the-art aircraft.

The challenge for any serious nation that invests in an Air Force is to select, train and employ the best combat air crews that they possibly can. You can train pilots simply with synthetic technology but an approximation of pilot effectiveness can not be made until real combat becomes the final and ultimate judge.

Techniques for transforming fledging students into proficient combat pilots have evolved through the years as the result of much research and development. Although training techniques constitute a necessary, although not completely sufficient, component, they are becoming increasingly important as weapons and warfare become more complex.

There are, of course contributors to pilot proficiency other than training techniques. The inborn abilities some pilots seem to possess play a huge part. But there is little reason to believe individuals with these natural abilities exists disproportionately among nations.

In fact, the actual combat history of kill ratios show that many nations can produce both Aces (5 kills) and even super-aces with many, many aerial victories. What clearly does play a role and can differ significantly from one nation to another are the cultural and social qualities that give aircrews the motivation to fight and the basic capacity successfully to use the technology in the aircraft and weapons they fight with.

"Flying should be an inherently dangerous business to weed out the weak sticks" is a Marine pilot's saying. One would hope that there could be a less dramatic and much more cost-effective method for developing aviators.

Understanding the cascading progression of how a nation that is serious in acknowledging the value of meritocracy in aerial combat selects and trains crews is simple to diagram, yet it is the unforgiving execution of rewarding excellence that means everything.

The cascading steps to produce a competent aircrew of an airpower-enabled fighting force that can prevail in any engagement and thus win wars can be broken down in phases to understand the progression of excellence in air combat:

- A nation's cultural factors that include simple motivation and advanced technology capability.
- A selection process that includes a rigorous human/technology match screening process within the consideration of the pool of candidates.

- Flight training that has to have a rigorous sorting process with a robust syllabus.
- Combat training that is based on a dynamic iterative training and readiness syllabus, the source of instructors and exposure to live firing and ordinance employment with appropriate use of simulators and ever improving dynamic tactics, along with demanding training ranges with enough airspace allotted to engage in multi-platform operations that incorporate both realistic air and ground threats.
- Proficiency training which if everything in the previous steps is solid and unforgiving and then can include war experience, or as necessary intense combat simulation events, a pilot's annual flying rate always with a command eye on the realism of aircrew training, and time in type.

It is in the early training toward their "wings" that all the worlds air forces must train their pilots to simply fly successfully, so that at least they will not crash their aircraft flying around the base flagpole. However, the real focus of creating successful squadron fighter pilot rests with the dynamics of combat training and then in subsequent continuous proficiency training as the individual rotates in and out of a squadron.

The list is not complete but combat training for the first tour "nugget" has drivers such as a Training and Readiness (T&R) syllabus. Different air forces have different names, but it is a check list of "hops" of increasing complexity that a newly arrived aviator must successfully accomplish to advice in sequence in order to become fully combat qualified.

A key intangible that should never be overlooked is the source of instructors during this combat training cycle, along with measurable indices such as live firing/weapons release, and simulator training. Combat training is a progression of building blocks through flying sorties of more and more demanding tactical and weapon training flights that will ultimately rise to the level to operate their fighter against the highest threat environment in the world.

Military technology is always relative against a reactive enemy and MAWTS-1 leadership and instructors know this perfectly. Modernization of military forces, especially air combat assets can usually be driven by three dynamics.

The first is to gain and successfully integrate new capabilities. For example, over the history of MAWTS the type/model/series (T/M/S) of aircraft progressed from the F-4 to the F/A-18 to the F-35.

The second is to add new components which provide for enhanced or more reliable operation of existing equipment, hence the ever improving both synergistically and independently developed weapons of a fighting air force.

The third is to simply replace worn-out equipment, often seen as advances in a developed T/M/S aircraft for example the decades long improvements from the F-4B to F-4S. The "block" improvements of the F-35 are following this modernization trajectory.

In aviation, basic airframe performance enhancements of payload, range, maneuverability, and speed to then enhancing aircraft system performance by incorporating a successful payload utility function of target acquisition to target engagement is the key at the squadron level.

However, in combat the total force performance capability goal is always to strive for excellence in having command and control systems that can tie it all together.

Honoring and empowering individuals engaged in the deadly serious occupation of defending their fellow citizens as combat warriors while putting their life on the line comes first before any future technology discussions can begin.

It is no good to talk about future technologies without starting from the nature of warfare and of human engagement in that warfare.

Often looking at ground battles from the earliest recorded days, the forces engaged had a simple guiding rule — kill the enemy in greater numbers. There is no hard and fast rule from history of what tips a battle one way or another except one core principle: with the will and means to continue to degrade one's opponent, winning is enhanced.

The great quip often credited to Grantland Rice, the famous 20th century sports journalist, who gives full credit to a fellow sportswriter comes to mind. As Hugh Keough, a famous Chicago sportswriter of the 20th century, used to say: "The race is not always to the swift, nor the battle to the strong; but that is the way to bet."

Such insights are actually biblical, drawn from The King James Bible (such poetic writing): "I returned, and saw under the sun, that the race is not to the swift, nor the battle to the strong, neither yet bread to the wise, nor yet riches to men of understanding, nor yet favor to men of skill; but time and chance happened to them all."

The key to combat success since the dawn of warfare is captured in a very simple example — the great command guidance in the USMC is to stress the very basic art of accurate marksmanship.[6]

"Ready on the Left Ready on The Right-Already on the firing line" and with that every Marine is trained in the use of their rifle.

Once trained, and retrained, and retrained until actual combat (because their skills are never allowed to atrophy), the individual Marine has in direct combat engagement using a very simple payload utility function in shooting the weapon. The combat utility of the basic rifle is acquiring the target and then accurately engaging to kill the enemy.

That type of engagement at the basic infantry level is no different than the senior Generals and Admirals having their fighting forces acquire and engage targets using many different mixed and matched payloads.

This universal way of war is often correctly referred to as combined arms, as layer after layer of direct and indirect fires, kinetic and non-kinetic, weapons are engaged to defeat the enemy.

In fighting against a reactive enemy in a larger battle, the aggregation and disaggregation of sensor and shooter platforms with actionable intelligence with no platform fighting alone is the commander's goal.

Commanders at the highest level must keep both cohesion of

the combat engagement mission by effective communications while concurrently relying on all to engage intelligently relying on their individual initiative to fight to the best of their ability.

Communicated information is essential. The key is to ensure a maximum of capability for combat operations to be able to operate independently with accurate real time dynamic intelligence at the right level at the right time to make their combat function superior to the enemy.

Very little is different from the deck of Navy Strike force or Air Battle or Ground Commander except the complexity of all the "moving parts" to be managed and employed to fight that are also spread out over very great distance.

At MAWTS-1, training of the trainers embraces that after two decades of the land wars, the sea services need to learn to fight again in higher intensity operations. It means to master the ability to fight at the speed of light. This requires that a fighting force at all levels must take advantages of ever-increasing technological advances to make decisions using the speed of light intelligence, sensors, and robust communication.

With advances in all forms of "tron" war, shorthand for information revolution, the moment battle begins, command and control are essential and has to have several attributes.

First and foremost, accurate information must flow through robust redundant systems at the speed of light in making everything come together to fight and win. The infantry platoon commander trusts the training and combat effectiveness of each Marine to do the right thing using initiative in following orders in the heat of battle while also trusting higher commands to provide supporting arms, including air, to get it right and at the right time. The sea service air battle commanders trust that the aircrews are fully trained and motivated.

Thus, communication and intelligence capability in this 21st Century evolution/revolution of global coms and sensing is the connective tissue for human decisions with how to conduct successful operations and to successfully engage payloads with all fighting at the speed of light.

America is blessed that many in the defense industrial base have answered the challenge to build systems of systems inside the emerging Kill Web way of fighting, vice obsolete Hub Spoke and linear Kill Chain thinking.

Existing command and control are always against a reactive enemy and there is a time dependent factor that is critical to force level combat. If a commander can count having the initiative with a combat ops tempo over the enemy, then his forces can be dynamically optimized as a coherent combat directed fighting force.

This is the challenge of effective command and control, but ultimately the commander has to always have the wisdom and judgment to fight to win effectively.

Marrying force motivation with technological capability allows a superior trained force to achieve combat performance over the enemy. It is a combination of appropriate combat equipment at all levels of any engagement operated by well-trained individuals. along with a satisfactory inventory of weapons systems and platforms, including sufficient munitions at the start of a war. Such preparedness can make all the difference.

The biggest challenge in the rapidly exploding human/information dynamic in this 21st Century challenge of modern war is the ability to make accurate decisions at light speed.

MAWTS-1 meets the challenge of driving home all the essential elements for achieving combat victory today, and tomorrow. This book is their story.

By Ed "Easy" Timperlake

1

The 2011 and 2012 Visits: The Coming of the F-35B

Our first two visits as a team to MAWTS-1 were on October 2011 and November 2012.

The interviews we published from those visits provided an overview on how MAWTS-1 was preparing for the arrival of the F-35B and how the team at MAWTS-1 was changing its approach to work with and master the aircraft – notably operating as a wolfpack – in order to dominate and prevail in the contested battlespace.

The F-35 is a very different aircraft. The Marines were first deploy the aircraft and are today the key fifth generation enabled force in the first island chain. That would not have happened without MAWTS-1 as our visits would underscore.

Preparing for the F-35B Transition: MAWTS Re-Shapes its Curriculum

October 20, 2011

We recently visited Marine Aviation Weapons and Tactics Squadron 1 to hear first-hand the effort to shape the F-35 transition. We also visited the construction team re-building Marine Corps Air

Station Yuma and will report on the changes being made to operate the F-35Bs in Arizona.

The days spent with MAWTS were invaluable in shaping an understanding of the impact of V-22 and F-35 on the changes in tactics and training generated by the new aircraft. MAWTS is taking a much older curriculum and adjusting it to the realities of the impact of the V-22 and the anticipated impacts of the F-35.

MAWTS is highly interactive with the various centers of excellence in shaping F-35 transition such as Nellis AFB, Eglin AFB, the Navy/Marine test community at Pax River Md, and with the United Kingdom. In fact, the advantage of having a common fleet will be to provide for significant advances in cross-service training and con-ops evolutions.

Additionally, the fact MAWTS is studying the way the USAF train's combat pilots to be effective flying the F-16 in shaping the Marine F-35B Training and Readiness Manual (T&R Manual) is a testimony to a joint service approach. This is all extremely important in how MAWTS is addressing the future.

General "Dog" Davis, the 2nd MAW Commander, had highlighted for us the importance of merging three USMC pilot cultures into one.

The CG emphasized the importance of the electronic warfare (EW) culture within the USMC, and to the Prowler community within the USMC.

The Prowler guys are some of the brightest guys in the USMC.

But the F-35B was going to provide the USMC aviator cultures in our Harriers, Hornets and Prowlers to coalesce and I think to shape an innovative new launch point for the USMC aviation community.

We are going to blend three outstanding communities. Each community has a slightly different approach to problem solving. You've got the expeditionary basing that the Harrier guys are bringing to you. You have the electronic warfare side of the equation and the high-end fight that the Prowler guys think about and the coms and jamming side of the equation, which the Prowler guys think about. And you have the multi-role approach of the F-18 guys.

I think it is going to be a fantastic blending of not only perspectives but also attitudes. And what I really look forward to is not the old guys like me, but the very young guys who will fly this fantastic new capability. The older generation may have a harder time unleashing the power and potential of the new gear – the new capabilities. We might say "why don't you do it this way" when that approach might be exactly the wrong thing to do from a capabilities standpoint.

My sense is the young guys will blend. We've already picked the first Prowler pilot to go be an F35 guy. He's going to do great and he's going to add perspective and attitude to the tribe down at Eglin getting ready to fly the jet that's going to make a big impact on the F35 community.

I think it's going to be the new generation, the newbies that are in the training command right now that are getting ready to go fly the F35, who are going to unleash the capabilities of this jet. They will say, "Hey, this is what the system will give me. Don't cap me; don't box me in. This is what this thing can do, this is how we can best employ the machine, its agility its sensors to support the guy on the ground, our MEU Commanders and our Combatant Commanders and this is what we should do with it to make it effective."[1]

At MAWTS, the historical curriculum has been built around modules which basically reflect core platforms or tasks. The V-22 is in the module with rotorcraft.

But the V-22 and F-35B do not fit in easily.

However, an emerging approach may well be to take functions and then to redesign the curriculum around those functions. For example, the inherent capabilities of the emerging F-35B cockpit with 360 situational awareness may turn out with appropriately designed data links to be a force multiplier in the tactical employment of the MV-22 Osprey and the helicopter community and reach back to Navy combat forces afloat. Everything is on the table to establish a true combat effective 21st Century "no platform fights alone" reality.

We discussed this transition with several Marines at MAWTS. The MAWTS understanding of the transition was outlined by four Marine officer we met with during our visit.

Major Derek Branon, head of the TACAIR department within MAWTS explained the nature of the transition from the F-18, AV-8B and EA-6B to the F-35B. Major Branon underscored that they were using scenario-based training as a means to re-shape the curriculum. The aircraft is very flexible so one can look at air-to-air, TRAP, CAS support and EW brought together in a new approach to con ops. The shift is from a target-oriented pilot culture to a mission or task-oriented pilot culture.

Major Joshua Smith, head of the MV-22 division within MAWTS explained the transition for the V-22 and how it operates differently for the USMC. Major Smith underscored how the experience with the V-22 transition is being looked at to shorten the lessons learned necessary to incorporate the F-35B into USMC mission sets.

The paradigm shift is underway with the deployment of the MV-22 and extrapolating from how it is currently working with the Harriers. He noted: *You will see exponential expansions of the distances and the capabilities you can take forward on the Amphibious Ready Group (ARG).*

Major Shoop is the F-35B lead at MAWTS in shaping the new training and tactics curriculum for the introduction of the aircraft into the WTI) Course. Major Shoop explained the nature of the transition associated with having an integrated sensor rich airplane. EW has been an external asset; now EW is brought into the cockpit of the tactical aircraft itself. This is a major change.

Major Kevin Crespo works with Major Shoop on developing and shaping the new training and tactics curriculum for the introduction of the F-35B into the WTI Course. Major Crespo focused on the ARG becoming a much more effective and scalable asset. He emphasized: *The new capabilities we are adding allow us to be able to keep the other side puzzling over what we are bringing ashore and where we insert that force and that provides us with a key strategic advantage.*

We concluded our visit by discussing the question "What impact will the first 6 F-35Bs aboard an ARG have on the first operations?"

There are obvious significant and revolutionary air-to-air implications, but much of our discussion as very appropriate for U.S. Marines revolved around the support to the 0311 fire team leader.

Among the many initial impacts are the following:

- An ability to suppress enemy communications.
- An ability to deal with all missile threats.
- An ability to operate essentially as a traffic cop to provide guidance on the most effective routes into the littoral where the adversary's forces are weakest.

Another Aspect of the F-35 Transition: Rolling Out New Infrastructure

November 1, 2011

The F-35 is the first significant build of new tactical aircraft in a long time. Correlated with the build is shaping a new infrastructure to support the new fleet. In many cases, the older infrastructure is outdated in and of itself and the new aircraft accompanies needed upgrades to infrastructure.

A new infrastructure is being built for the F-35. The new infrastructure is shaped to support the approach to the overall new weapon system, notably how the simulation centers are going to be used in training, and mission planning. New simulation centers are high technology, electronic centers of excellence that in turn need a proper power and communications infrastructure.

In addition, the F-35 with its C4ISR D capabilities will be maintained and operated out of secure hangers and facilitates. Securing the communications and support built around the ALIS system is a key element of the F-35, which will be protected in part by being located in secure facilities.

During our visit to MAWTS, we had an opportunity to talk with the SEABEE led construction team at Yuma Air Station. There was considerable enthusiasm in the team and progress was evident throughout the base.

Among the core elements for a new infrastructure being built at Yuma are the following: 2 Aircraft Maintenance Hangers, an Intermediate Maintenance Activity Facility, a Communications Infrastructure Upgrade, a Simulator Facility, and a Utilities Infrastructure Upgrade.

There is considerable cross-fertilization among the various infrastructure construction efforts. An advantage of having a global common fleet is that as new infrastructure is crafted, a lessons learned process is generated whereby next iterations of an infrastructure roll out become progressively more cost effective.

As one participant in the roundtable on Yuma construction underscored:

There is a branch in the USMC for lessons learned. And each of these guys here has put together point papers for lessons learned. We are in the process of disseminating our lessons learned through all of the services.

This is the advantage of this being born joint. As we have rolled out our first hanger, we have learned about how to build these new secure facilities. And we are in constant dialogue with Eglin regarding their experience as well.

One of the more interesting observations we can offer is that any delay in rolling out the infrastructure has been caused by the U.S. government itself and the constant starts and stops in cash flow. As one participant noted:

It is hard to build multiple parts of the infrastructure with contracted construction companies and then you have to idle them and then bring them back on at the same time. They can not work on top of one another.

In the inside the beltway debate, folks have lost sight of the advantages of fleet commonality within the manufacturing process and seem to have no clue as the very significant capabilities and savings inherent in a "lessons learned" process for building out the infrastructure to support a common fleet.

The F-35 Squadron at Yuma: The Next Phase Begins

November 18, 2012

Later this month the first two F-35 Bravos are arriving at Marine Corps Air Station (MCAS) Yuma.

As the future home of five F-35 JSF operational squadrons of 16 aircraft each and one operational test and evaluation squadron of eight aircraft totaling 88 aircraft, these squadrons will replace Yuma's four existing squadrons consisting of 56 AV-8B Harriers.

By early next year, the full complement of 16 F-35 Bs will have arrived. For the USMC, this is the beginning of the next 100 years of naval aviation.

It is also the ultimate response to the terrorists who blew up 1/10 of the Harrier force. The USMC will give them back and powerhouse combination of the Harrier, the F-18 and the Prowler with new C2 and Information Warfare capabilities.

Here they are following the model of the Osprey roll out. Once the aircraft was ready to fly and to be part of a training effort, the Marines began to use it. They understood that the capabilities of the aircraft would be rolled out over time, and further aspects of that capability evolved over time. They also understood with the Osprey that determining how best to use the aircraft and how those operations which affect overall Marine Corps operations had to be determined in practice, not in an abstract, linear evaluation process.

This is why the Osprey at the 5-year deployment mark is essentially seen as a different aircraft in terms of how the Marines can and do use the aircraft. They have also learned how the physical aspects of the platform affect operations, and, in turn, shape concepts of operations. Only with real world experience in Iraq, Afghanistan, off the Shores of Tripoli, or in exercises such as Bold Alligator 2012, would the evolution of the aircraft be shaped.

Although an even more significant addition to the Marine Corps and its con-ops, the F-35 Bravo is being approached in a similar manner. As the baseline aircraft becomes capable of entering the training and squadron evaluation process, the Marines can then determine how best to evolve the core capabilities of the aircraft.

As Col "Turbo" Tomassetti, the Deputy 33rd Fighter Wing commander has put it:

Once we have the basic aircraft flying and the core operational capabilities enabled, we need to shift from a block development process determined by engineers and have an operators determination of next steps needed. This is especially true given the unique fleet qualities of forthcoming F-35 operations.[2]

While the squadron is being established at Yuma, the USAF will begin taking planes at Hill Air Force Base and starting to put together their first F-35 squadrons. And during all of this training at Eglin, and further evaluations at Edwards AFB and Pax River Naval Air Station will continue.

After initial deployments, the aircraft will evolve as operators determine best approaches to getting incremental yet significant combat value out of the aircraft. And this is why the collocation of the squadron with MAWTS is so crucial. MAWTS is where the Marines develop tactics and training for the various aviation assets working with overall Marine Corps operations. At MAWTS, the Marines shape their approach to innovation as they move forward, notably with new systems, or newly configured systems.

The squadron at Yuma will shake down the aircraft and get it operational. As they do so, the pilots using the plane will work closely with MAWTS in shaping the new tactics and training associated with the aircraft.

Because this aircraft is a bundle of Harrier, Prowler, and F-18 capabilities with its own revolutionary foundation to doing air operations, the impact of using the aircraft will be central to the evolution of tactics and training.

Notably, Prowler pilots have been added to the MAWTS team in preparing for the F-35. As one MAWTS instructor put it: *Prowler pilots are information warriors, and this is a core element of what the F-35 is all about.*

With MAWTS working closely with the squadron, the development of tactics and training which are an inherent part of development for the plane, the squadron and the program, will be a center bull effort. And this will be significant as the squadron moves out. Already, the deployment of the Yuma squadron to Japan is envisaged.

And in an interview with Lt. General Jan-Marc Jouas, the 7th USAF commander, the significant role which the F-35Bs can play in his mission in the defense of South Korea and to provide for greater U.S. combat capability in the region have been underscored.

U.S. overseas basing decisions are not yet determined; however, any deployment of F-35s to the Korean peninsula will clearly modify the template, including the Marine Corps F-35B.

MAWTS-1

The Seventh Air Force relationship with the Marine Corps is the best I've ever seen. Their aircraft will be dedicated to the Marine Air Ground Task Force (MAGTF) at some point, but before then, they will be used as part of our air campaign to the greatest effect that we can deliver.

The F-35A, B, and C will give us greater flexibility, and greater options in terms of where and how we can operate.

We will integrate the F-35 with F-16s, F-15Ks, F-15Es, F-22s, and other airplanes in a way that will enhance and increase everybody's capability, much in the same way that we currently see the F-22 and the F-15 integrating and increasing their capabilities. Our targeting, and the effects that we will seek, will be adjusted by the fact that we have F-35s.[3]

In other words, the F-35 is a key asset in shaping the "Pivot to the Pacific." It is a lynchpin program in a lynchpin strategy. In addition, the USS America will be home ported in San Diego and empowered with up to 23 F-35Bs on its decks, dependent upon the mission configuration of aircraft on its deck.

As Captain Hall, the prospective commander of the USS America recently underscored:

We are a large deck amphibious ship, just as the Kearsarge. But we are an aviation-centric large deck amphibious ship and we've been designed specifically without a well deck so we can support the USMC's next generation of aircraft.

We can get out there with a much larger hanger bay with two high-hat areas to support maintenance on the much larger MV-22s. The maintenance requirements for the F-35 are met and we have the capability to expand when required for future development. With our added fuel, ordnance, maintenance capability, supply and support capacity, we can sustain the aviation capability much longer on station.[4]

And as Major General Walsh emphasized, the F-35s flying off of the deck of the USS America is not just a generational leap but a quantum leap.

When I went from flying F-4s to F-18s that was a shift. With the F-35 it is a leap of multiple generations all at once. It's more of exponential curve than we did when we went from third generation to fourth generation. It will be bringing electronic attack and C5ISR to the USS America as a presence asset. This will be revolutionary. Putting the new aviation assets together with the new ship will create the possibility of having a "MAGTF on steroids".[5]

The twin transition for the USN-USMC team is highlighted on the USS WASP deck as seen by Robbin Laird on October 18, 2011.

And the experience at Yuma and the USAF at Hill be replicated throughout the global fleet of F-35s. A core advantage of the aircraft is that as a global fleet its support structure has significant commonality which allows for cost savings, and more effective collaboration among the services and the allies.

This next year already the British and Dutch will be at Eglin AFB for training. And the Australians, the Japanese, Italians, and Norwegians are already on board in procuring the aircraft. Getting that common experience from the initial squadrons will be a core element of deploying the aircraft and shaping the combat capabilities of the forces using that aircraft.

As LtGen Robling, the highest-ranking Marine in the Pacific has put it: *The challenge facing the USN-USMC team in the Pacific is persistent presence. And the F-35 operating on the joint and coalition level will be essential to the way ahead in executing such presence. How can the allied F-35s work with yours to shape a new Pacific capability?*

First, we would have common or like support structures. This will increase are forward readiness posture by being able to fix and maintain aircraft that are deployed vice send them back to the states for repair or reach back to the U.S. for parts. The more allies who buy the aircraft the more spread out that support structure would be.

Second, the capability of the aircraft as a C5ISR platform will allow significant sharing information sharing and fusion to more of our partners who are able to receive and use the information. This increases our persistent presence capability. The aircraft will help fill in capability gaps or seams between us and our partnering countries and in the end, help build or increase their own capacity.

The F-35B squadron at Yuma in November 2012 begins the process. And cumulative U.S. and allied experiences will build out the development process the aircraft and the fleet in action. It is the beginning of the next phase of the F-35 program, equivalent to what the USMC did 5 years ago in Iraq with the Osprey.

The future does not belong to the timid, but to those whose lives depend on getting 21st century capabilities into the combat force. Enhanced persistent presence for a 21st century strategy rolls out one F-35 squadron at a time.

Leveraging the F-35: MAWTS Prepares the Pilot of the Future

November 25, 2012

Because the F-35 is a bundle of Harrier, Prowler, and F-18 capabilities with its own revolutionary foundation to doing air operations, the impact of using the aircraft will be central to the evolution of tactics and training.

In a discussion with Major Clint "Boo Boo" Weber, the Head of Tac Air at MAWTS, and with his colleague Captain Roger "Hazmat" Greenwood, the fit between the preparation of the aircraft for Marine Corps operations with the aviators was discussed.

Question: Could you discuss the process of re-alignment of MAWTS personnel with the coming of the F-35?

Weber: *When we discuss MAWTS-1 personnel to be involved with the JSF, we have to go back to when General Trautman was Deputy Commandant of Aviation. It was determined at that time that it was necessary to ensure that MAWTS-1 was essentially at the forefront of operational test when it became appropriate.*

And every DCA since that time has re-affirmed this approach. It has been felt that operational tests would be concurrent with tactics evaluation for the new jet.

As we have gone through this process, we have tried to get the right guy for the right job. "Hazmat" brings significant background to MAWTS, which is crucial for where we are going. He has significant experience operating F-18s off of carriers and is a graduate of the USN's Fighter Weapons School.

He was an instructor at Top Gun with regard to tactics and training. His experience in the joint environment and operating at sea is a key part of where we are going in the future.

After MAWTS, he will go to Edwards to work on weapons integration and fusion integration. His background and experience are crucial for this next assignment as well and of course he will feed that back to us here at MAWTS as well.

"Hazmat" Greenwood: *I think the operational experience with the Navy will be important as the Marines insert F-35s into the force. The Marines are in an interesting spot, as we will have the plane first and can provide some insights into how the tactics and operational concepts will change with the plane. We can provide inputs to our Navy brethren with regard to these developments. We will be leading forward on the impact of the F-35 transition for our sister services.*

General Amos on the F-35s at Yuma

November 29, 2012

The Marines have stood up their first squadron of F-35 Bs at MCAS Yuma. But the Marine Corps approach to the aircraft is built on recognition that it is a C2 and Information Warfare aircraft, which will be a central piece to the ACE or Aviation Combat Element of the MAGTF.

During the re-designation ceremony, General Amos, the USMC Commandant, highlighted the nature of change and the role of the F-35 in this process.

The F-35B is the future of Marine tactical fixed wing aviation. As many of you know, today the F-35 begins replacing three models of tactical jets the Marine Corps currently operates. In fact, VMFA(AW)-121 gave up their F/A-

18D Hornets just a short three months ago after returning from a highly successful WESTPAC deployment.

The F-35 will replace our Hornets, our AV-8B Harrier attack aircraft and our EA-6B electronic warfare aircraft.

Replacing so many different platforms with a single, multi-capable aircraft represents a new way of operating and thinking. This jet possesses "eye-watering" capabilities.

The things it can do are most impressive to a couple of old F-4 Phantom guys like General John Hudson… and myself. Unfortunately, I can't talk about most of those capabilities here.

But, suffice it to say this is not your father's fighter!

VMFA-121 is at the forefront of one of the most significant transition periods in the 100-year history of Marine Aviation, as we replace nearly every aircraft in the Corps between 2005 and 2025.

Certainly, it is the most significant transition in quite some time, maybe since the introduction of the helicopter to our forces in the post-World War II 1940s. But, being on the forefront is not new to this squadron, the "Green Knights" have been a storied squadron since they were established just months before the attack on Pearl Harbor…..

I noted earlier that having F-35Bs in Yuma shows tangible progress in this vitally important aircraft program.

There is additional progress all over our Corps today regarding fielding the F35B. I call your attention to the seven F-35Bs currently at NAS Patuxent River, Maryland, conducting flight test activities, the 11 F-35Bs now at our training squadron, VMFAT-501, at Eglin AFB, Florida and the two United Kingdom F-35Bs that have joined 501 and have also begun training there.

We are making strides in every aspect of this program. Aircraft are being produced tested and flown, pilots are being trained in the air and in the simulators and aircraft mechanics and technicians are learning to ply their trade on this magnificent jet. Yuma will eventually have five operational squadrons and be responsible for operational evaluation of the F35B. MCAS Yuma will continue to be a busy base.

MAWTS and the Yuma F-35 Squadron: Evolving Capability into Operational Reality

November 29, 2012

The squadron at Yuma will shake down the aircraft and get it operational. As they do so, the pilots using the plane will work closely with MAWTS in shaping the new tactics and training associated with the aircraft.

In a discussion with Major Clint "Boo Boo" Weber, the head of tac air at MAWTS, we discussed the Marine Corps approach. Weber is Tac Air Department head at MAWTS, which means he works with all the fixed wing aircraft used by the USMC, which would include F-18s, Harriers, EA-6Bs, KC-130s and UASs.

Question: How does MAWTS look at the deployment of F-35s at Yuma?

Weber: *VMFA-121, the Green Knights has stood up already, but will receive its first aircraft on November 20th. And we'll continue to receive aircraft at a substantial rate such that they should be fully stood up by next spring with 16 total airplanes.*

Question: What will be the approach of MAWTS to the operational squadron?

Weber: *We're hopeful that by having the squadron right out here at Yuma next to our weapons school, Marine/Aviation Weapons and Tactics Squadron 1 that we can work on concurrent tactical and operational development. And we are part of the conversation on concurrent developmental tests and operational tests at Pax River and Edwards and Eglin and then at China Lake.*

But as that goes on concurrently, we are flying and developing tactics for current software and current capability with fleet aircraft with the aid of MAWTS-1 instructors here.

For us to be right next to them, that's going to be extremely important when we start talking about tactics evaluation for that squadron that deploys in the next two years.

MAWTS-1

Major "Boo Boo" Weber, Head of Tac Air at MAWTS, with Ed Timperlake during our 2012 visit to MAWTS-1.

Question: Talk about the impacts, which you see from the F-35 on Marine Corps tactics and training?

Weber: That's really a key point. What I see as really our primary responsibility here is we start taking a look at initial tactics evaluation or development of tactics.

In the Marine Corps, we're looking to declare Initial Operational Capability (IOC) for this airplane get it airborne in numbers with its current capability in an environment where we can start to work it into the game plan. Without using the aircraft, it is impossible to develop the aircraft into its operational capabilities and determine its overall impacts on Marine Corps operations.

The Commandant has emphasized the return to the sea. The Bravo is a centerpiece of the kind of at sea capability which is central to the Marines. With more difficult environments to operate in, the F-35B is part of assuring of us of greater capability to operate in a variety of settings.

The Osprey is also a part of this, but the Bravo will not only support operations but be a hammer to knock open holes in difficult operational environments. It is also enables us to operate in a much wider range of environments. The

Bravo can operate off of a variety of surfaces -- ships, airfields, highways, and fields.

The Marines have specialized in setting up airfields where there aren't any. We think this is a core competence, which will be in high demand in the future. For the U.S. Marine Corps, the F-35B gives us more flexibility on where we can fly it from, which is probably the most important part for us.

Question: We noticed that you have a broad range of pilots feeding into the F-35. You have Harrier, F-18, and Prowler pilots, to name three. How is that going to shape the culture of the aircraft?

Weber: There's no doubt that initially when we start talking about the syllabus of the communities that are primarily going to feed into the JSF that you will see change.

You have the fighter attack syllabus and the attack syllabus for the F-18 community and then in the AV8 community. And you need that because you definitely need the pointy-nose guy out there. You're going to have a high-end platform, fifth generation aircraft that's going to have some capabilities that are unmatched.

And there's really only one type of individual that can take advantage of that. That's somebody from the fighter and/or the attack community.

But there's some other considerations as well when discussing the F-35. It is a C2 and Information Warfare aircraft. Our best operators in this world are from the Prowler community, and they are key to shaping the F-35 culture.

For the Marines, such integration is crucial. We are naval officers, we are Marine Corps officers, but above all we are MAGTF officers. Which means that once you become a bit more mature, you have to start thinking about not just airspace or battlefields but battlespace. In other words, it's important that the individual that's building the syllabus for the F-35 thinks about command-and-control.

Whether you call it cyberspace or information warfare or pushing the right pieces of information to the right places, the F-35 is a centerpiece for Marine Corps thinking about the future.

In other words, you don't want to put people in charge of training, creating training syllabi, or essentially, creating tactics that are thinking simply at the air operations level. You have to think about this holistically from the MAGTF perspective and not just operationally from the air or from the air for the air.

We have to think operationally from the air for all the other elements of the MAGTF. The ground combat element, the logistics combat element, and then how that works into the joint arena as well are crucial when we think about tac air in the USMC and at MAWTS.

2

A 2013 Perspective: The Yuma Incubator of Change

In our 2013 book written with Richard Weitz, we focused on how to rebuild American military power in the Pacific. Even though there was an announced "Pivot to the Pacific," the priority remained on the land wars.

And several years of continued engagement meant that investments in systems for what later would be termed "the Great Power competition" were not made. These land wars would be guillotined by President Biden's Blitzkrieg withdrawal from Afghanistan.

This meant that the United States had to find ways to leverage the new systems it was building and deploying to provide for coverage of the Pacific. This would later lead to a significant shift towards operating and then enhancing a distributed force. For the Air Force this was agile combat employment. For the Navy this was distributed maritime operations. For the Army it was an adventure.

The Marines are a force built to distribute and to operate with flexibility, something the other services simply have not been built to do.

We argued in the book that the Marines could lead the way in the Pacific force transformation, with their Osprey, with their lead role in deploying the F-35, and with the coming of their new heavy lift helicopter, the CH-53K.

And where would leadership be provided for this effort?

We argued that MAWTS-1 would have a special role, notably in working with her sister training centers. the Air Warfare Center for the USAF and NAWDC for the U.S. Navy. We focused on these other centers as well as the services worked to shape a new agenda and new concept of operations for a joint force able to compete with peer competitors rather than simply supporting the U.S. Army in it Middle Eastern engagements.

We argued in our introduction to that book: *In short, the challenges are significant—the rise of China, the North Korean threat, the emergence of the Arctic, and the challenges associated with counterterrorism and maritime security.*

At the heart of an effective response will be shaping innovative relationships between the United States and its allies and coming to terms with ways to deflect Chinese expansion while at the same time working with China in shaping global prosperity.

The challenge will be to forge effective building blocks through partnerships, technologies, and organizational innovations that can provide a 21st century of security and defense in the Pacific. A key element for success is training and preparation for the 21st century, not remaining in the mind-set of the 20th.

One U.S. Marine Corps (USMC) general referred to the need to shape new capabilities for the I-Pad-generation pilots, not for his own generation.

What he had in mind was the touchscreen cockpit of the F-35 and its ability to work with the visual acuity of the new generations. It is always important to remember that the human element in military operations is the technical skill, resourcefulness, and dynamic innovation of members of the fighting force combined with individual courage, training, and competent leadership at all levels —and often, as Napoleon said he liked to have, a lucky general.[1]

In that book, we highlighted what we saw as a key role for MAWTS-1 in helping shape a way ahead in the Pacific. What we wrote then is what follows in the rest of this chapter.

The F-35B Comes to Yuma

Far from Washington and its think tanks, Marines and their sister services are engaged with them in shaping new con-ops for the future. Battle-hardened Marines are at the heart of forging the future.

The Marines have stood up their first squadron of F-35 Bs at MCAS Yuma. But the Marine Corps approach to the aircraft is built on recognition that in addition to its role as a strike aircraft, it has C2 and information warfare capabilities, which will make it a central piece to the ACE or Aviation Combat Element of the MAGTF.

Two squadrons are being established and are the operators. MAWTS-1 will develop tactics and training for the F-35 B in conjunction with the other aviation elements for the ACE.

And VMX-22 will focus on the technologies and systems of the platforms making up the evolving ACE for the MAGTF. The colocation of VMX-22, the F-35 squadrons and MAWTS-1 will facilitate the kind of culture change, which the F-35 enables for the MAGTF.

As Michael Orr, the CO of the squadron underscored: *We are testing all USMC platforms working together in developing Marine aviation capabilities. We are not just testing individual platforms. This is especially crucial when you have dynamic, transformational platforms such as the Osprey and the F-35.*

The F-35Bs have started their service life at Yuma Marine Corps Air Station. As the future home of five F-35 JSF operational squadrons of 16 aircraft each and one operational test and evaluation squadron of eight aircraft-totaling 88 aircraft, these squadrons will replace Yuma's four existing squadrons consisting of 56 AV-8B Harriers.

By 2013, the full complement of 16 F-35 Bs will have arrived. Here they are following the model of the Osprey roll out. Once the aircraft was ready to fly and to be part of a training effort, the Marines began to use it.

They understood that the capabilities of the aircraft would be rolled out over time, and further aspects of that capability evolved over time. They also understood with the Osprey that determining how best to use the aircraft and how those operations which affect overall Marine Corps operations had to be determined in practice, not in an abstract, linear evaluation process.

This is why the Osprey at the 5-year deployment mark is essentially seen as a different aircraft in terms of how the Marines can and do use the aircraft. They have also learned how the physical aspects of the platform affect operations, and, in turn, shape concepts of operations. Only with real world experience in Iraq, Afghanistan, off the Shores of Tripoli, or in exercises such as Bold Alligator 2012, would the evolution of the aircraft be shaped.

Although an even more significant addition to the Marine Corps and its con-ops, the F-35B is being approached in a similar manner. As the baseline aircraft becomes capable of entering the training and squadron evaluation process, the Marines can then determine how best to evolve the core capabilities of the aircraft.

While the squadron is being established at Yuma, the USAF will begin taking planes at Hill Air Force Base and starting to put together their first F-35 squadrons.

And during all of this training at Eglin, further evaluations at Edwards AFB and Pax River Naval Air Station will continue.

This process underscores a basic reality of the F-35 as a combat system. Rather than looking at the evolution of the program in Block steps, it is better to understand it in terms of operational clusters.

To date, the software and systems of the F-35 to fly the aircraft have been steadily put in place. Then the core combat systems are being plugged into the software upgradeable aircraft. The weapons are being certified at Edwards AFB to add strike and defense capabilities to the aircraft. Then the aircraft is ready to deploy.

After initial deployments, the aircraft will evolve as operators determine best approaches to getting incremental yet significant combat value out of the aircraft. And this is why the collocation of the squadron with MAWTS is so crucial. The Marines integrating aviation into overall operations is the core operational reality for them. At MAWTS, the Marines shape their approach to innovation as they move forward, notably with new systems, or newly configured systems.

The squadron at Yuma will shake down the aircraft and get it operational. As they do so, the pilots using the plane will work closely with MAWTS in shaping the new tactics and training associated with the aircraft. With MAWTS working closely with the squadron, the development of tactics and training which are an inherent part of development for the plane, the squadron and the program, will be a center bull effort.

For the Marines, the linkage of the F-35 with the Osprey and other air and ground elements is re-shaping their basic approach and combat power of the MAGTF.

The Perspective of Lt. General (Retired) Trautman

In a discussion with the former Deputy Commandant of Aviation, Lieutenant General George Trautman, this evolution was discussed. It is important to note that several years prior to the recent Yuma events, the USMC leadership had set in motion ways to leverage the new technologies.

In other words, leadership matters in shaping the cultural revolution associated with the Osprey and the F-35. The revolution is not concomitant with simply buying and introducing new technologies or platforms but is rooted in changing operational behavior and thinking, which the new technologies allow. And reciprocally, such changed behavior will shape the evolution of the platforms and technologies.

This will be especially true of a software upgradeable aircraft like the F-35. The nature of the cultural change associated with how the USMC approaches the F-35 and the Osprey was highlighted by the former Deputy Commandant of Aviation, Lt. General (retired) George Trautman.

After the ceremonies highlighting the arrival of the F-35B at Yuma, we discussed with Trautman his perspective on the approach to change. As the architect of putting in place the triangular working effort among the squadron, MAWTS and VMX-22 he was an ideal person to discussion the approach to change.

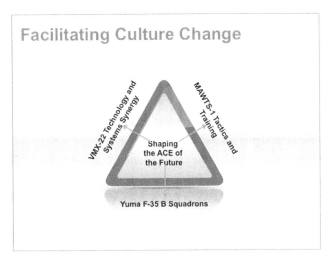

This is how we conceived of the impact of MAWTS-1 in the strategic shift in warfighting con-ops facilitated by new systems. Credit: Second Line of Defense

QUESTION: We would like to highlight something we find very innovative by the Marines, which was set in motion during your time as Deputy Commandant of Aviation (DCA). You are not just going to have a couple squadrons at Yuma. You've got the intersection with MAWTS and also the Marines are moving VMX-22 from New River.

In other words, the Marines have collocated three key elements to drive real world combat innovation associated with the aircraft.

Could you talk a little bit about the thinking that went behind setting that up and what your hopes are about that kind of collocation driving innovation?

Trautman: *As I look back to the summer of 2007 when I became the Deputy Commandant for Aviation, I marvel at the prescient decisions that our F-35B planners made in the run up to our first operational squadron.*

As you recall from previous conversations, one of the smarter things that we

did is we took Lieutenant Colonel Chip Berke and we assigned him to an F-22 squadron at the invitation of the Chief of Staff of the Air Force.

We put Chip there for almost three years and then we moved him down to take command of our first F-35 replacement training squadron, VMFAT-501. His experience with 5th generation operations in the F-22 informed us that the F-35 was going to require a significant adjustment in the way we train and fight our tactical fixed wing aircraft.

We knew from the beginning that we needed to focus intently on what it would take to ensure success when our first operational F-35 squadrons began to receive this game changing technology.

Of course, we handpicked the first operational squadron's new commander along with his initial cadre of aviators, some of whom who had previous experience with F-35 operations in VMFAT-501, in order to ensure we had extremely talented individuals manning VMFA-121.

But a key component of our decision to start out in Yuma was driven by the fact Marine Aviation Weapons and Tactic Squadron One, the world's premier organization for the development and employment of aviation weapons and tactics, is co-located on that base. MAWTS-1 is staffed with individuals of superior aeronautical and tactical expertise who are subject matter experts in every element of the Marine air/ground task force.

In my view, some of the greatest minds in modern aviation reside in that squadron. The commander is one of the best thinkers at the colonel level in the Corps today and his team has been charged by the current Deputy Commandant for Aviation to work with VMFA-121 to speed the development of future tactics and standardization in the F-35.

These two squadrons, operating side-by-side at MCAS Yuma, are going to reap incredible dividends for Marine aviation.

But the innovative approach to posturing F-35 for success didn't stop there. The Marine Corps also decided to take VMX-22, our only Operational Test and Evaluation Squadron, which had previously focused solely on the V-22 and expand its operational test portfolio to include the F-35B at MCAS Yuma.

With an F/A-18 pilot now commanding the squadron, the knowledge and lessons learned in operational test of the F-35B will inform the MAWTS-1 and VMFA-121 tactical planners as they stay focus on minimizing risk and maximizing performance of the F-35B in support of the Marine Air Ground Task Force.

With squadron operators, tactical innovators and operational testers all working toward the same desired end state at MCAS Yuma, this is going to be an example of a case where the sum is greater than the collection of its parts.

We just have to watch with a little bit of patience over the next year or two and I think you'll see that we'll reap the benefits of this decision far more than we would have if these organizations were operating at disparate geographic locations.

QUESTION: How will this kind of cross cutting innovation among the squadron, MAWTS and VMX-22 shape change?

Trautman: *Your premise that 'we don't know what we don't know' about this game changer called "fifth generation" is right on the mark. When young aviators, young maintainers, and young logisticians start to operate F-35 with the other elements of the MAGTF, we're going to experience exactly what we experienced with the V-22 when they first got their hands on the Osprey. There will be an exponential increase in the innovation and thinking and utilization of these platforms in ways that the initial planners who set the program's course never thought of before.*

In fact, I anticipate F-35 sparking a 5th generation intellectual engine at Yuma that will exceed all expectations.

An added benefit, of course, will occur when the Air Force stands up their F-35 squadrons at nearby Luke Air Force Base and we both start to operate F-35's on the ranges at Nellis. I'll make a prediction — three years from now, no one is going to want to be caught dead or alive operating a legacy jet in battle space dominated by the F-35 Lightening II.

When I look at Marine Aviation, I'm extremely pleased with where we are today. The V-22 Osprey does things that no airplane in the history of mankind has ever been able to do with regard to range, speed, maneuverability, and aircraft survivability — all combined with the ability to still conduct vertical landings in the objective area. The F-35B, a short takeoff and vertical landing machine with low observable characteristics, is a flying sensor that can do everything that we need it to do in ways that are simply going to change the game.

Having those two platforms side-by-side while the rest Marine aviation forms a complementary role and determining how all the pieces of aviation fit together is going to be a challenge, but it's the kind of challenge that I think all Marines relish and embrace.

This is what professional aviators do. This is what Marines do. It's what warriors in the other services do as well and, with V-22 and F-35, we are definitely providing the tools our Marines need to move these capabilities into the next century.

VMX-22 and the Yuma Incubator

The colocation of VMX-22, the F-35 squadrons and MAWTs will facilitate the kind of culture change which the F-35 enables for the MAGTF.

In interviews with the VMX-22 Squadron Commander, Col Michael Orr, he discussed with us the role of VMX-22 in shaping the future con-ops approaches. Orr describes the basic role of the squadron and its evolution as it transitions to Yuma.

QUESTION: Could you explain the role of the squadron?

Orr: *I'm privileged to lead Marine Operational Test and Evaluation Squadron 22. We conduct operational testing on assigned USMC aircraft under the authority of the Commander, Operational Test and Evaluation Force and the direction of the Deputy Commandant, Aviation.*

We used to be called Marine Tilt-Rotor Operational Test and Evaluation Squadron 22, and the name of the squadron and the number of the squadron should give away a little bit about its lineage. The squadron was created during the development of the V-22 Osprey program because of the services' emphasis on the successful operational evaluation of the V-22.

QUESTION: So the squadron was a crucial element of getting the V-22 out to Iraq and then on to Afghanistan?

Orr: *Absolutely. In order to reach full rate production and get the aircraft to the deploying squadrons, the V-22 needed to pass its operational evaluation.*

The Marine Corps' leadership at the time was not satisfied with the test structure that was in place and wanted to formally establish its own operational test and evaluation squadron.

This was a big step for the Marine Corps.

Major General Walters, at the time Colonel "Bluto" Walters, was the first CO. His task was to get the V-22 through its formal operational evaluation period and get it out to the hands of war fighter.

MAWTS-1

The squadron was remarkably successful in that role, built and trained the first cadre of operators and essentially laid the groundwork for the V-22 to go on and be a very successful fleet aircraft.

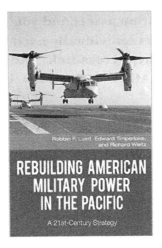

We put the first landing of the Osprey on a large deck U.S. Navy carrier on the cover of our book and that Osprey was piloted by Orr.

QUESTION: Prior to this the Marines tested through a Navy structure?

Orr: *For the most part we did. Operational testing was conducted by HMX-1 for assault support aircraft, VX-9 in China Lake for our fixed wing and H-1 aircraft, and VX-1 at Patuxent River for C-130s.*

After the launch of the Osprey, the squadron's focus was on testing upgrades on the V-22 and on some new mission systems.

A lot of that changed under Lieutenant General Trautman, then Deputy Commandant of Aviation, when he made a decision a few years ago to expand the scope of the squadron to include not just the V-22 and the CH-53, but also the F-35B and future Marine Aviation platforms.

I'm pretty sure that's why a fixed wing guy like me was chosen to head this new organization. I'm an F/A-18 driver by trade and I had trouble spelling the word "test."

I'm still learning a lot about the test mission, and I have tremendous respect for the professionals who know this business inside and out. My primary focus is to lead this squadron in its transformation from a tiltrotor and rotorcraft squadron to an ACE or Aviation Combat Element-focused test and evaluation squadron for the MAGTF.

QUESTION: In effect what you're doing is you're taking the new aviation assets and you're figuring out how those aviation assets individually need to operate, operate as a fleet, and then cross-link and obviously with a distinct eye to MAGTF operations because you're all MAGTF officers.

Orr: *Developing this interoperability of aircraft and systems is where we will see revolutionary improvements in the capabilities of Marine aircraft. It is an exciting time to be involved in Marine Aviation as all these new platforms come online in the next few years.*

How do we get the aircraft to work together? Who currently focuses on all the systems that are going to be required to achieve true interoperability?

Right now the squadron is located out of New River, North Carolina, which made sense when you were a tilt-rotor and rotorcraft-based test squadron, but we will be moving to Yuma in the next few years, which will be critical re-focusing our efforts on the entire MAGTF.

When we're in Yuma we're going to be set up right across the street from MAWTS 1, and we're very excited about this because this will give us the opportunity to have systems development work side-by-side with tactics development.

And if you can combine those two elements, you've got a very powerful tool that the Marine Corps can use to drive innovative change for the future.

Our participation in F-35B testing has two focus areas — which complement each other to bring warfighting capability to the Combatant commanders as rapidly as possible.

Our first focus area is to fully participate and support the Joint Operational Test Team activities, which are we going to be centered out of Edwards AFB.

Our second focus area is to begin to tackle the MAGTF integration piece of the F-35.

QUESTION: So the colocation of the F-35 squadrons with MAWTS with VMX-22 is designed to facilitate the culture of change associated with the Osprey and the F-35?

Orr: It is. *By bringing together the operators' perspective from the fleet squadrons, the tactics and training focus of MAWTS-1, and the systems development focus of VMX-22, we create a powerful combination to drive innovation for MAGTF aviation.*

We are still learning how the F-35 will reshape MAGTF operations so getting these three commands working together will help shape future concepts of operations and employment.

3

The 2014 Visit

2014 was a year where the Marines were spearheading the "Pivot to the Pacific." They were doing so without enhanced funding and relying largely on their Osprey aircraft to demonstrate the reach and range the U.S. forces would need to operate in the region. The only problem is that the Obama Administration was actually ramping up its engagement in the land wars, with its priority on Afghanistan.

But the Marines were getting closer to introducing the F-35B into their operational forces and the first to do so within the U.S. joint forces. This meant that as the Marines were preparing to fight in the land wars, a part of the force was being prepared to operate in the Pacific.

We visited MAWTS-1 on July 16 and 17 2014 and got updates on the F-35 and the Osprey and their way ahead for the USMC. At the same time. we were visiting 2nd Marine Air Wing and focusing on the new large deck amphibious ship, the USS America. The USMC was being transformed by new at sea assets and new air capabilities. MAWTS-1 needed to focus on the evolving integration challenges.

MAWTS-1 and Shaping the Future of USMC Aviation Within the Marine Corps

August 4, 2014

During our recent visit to Yuma Marine Corps Air Station, we talked with officers from both the VFMA-121 Squadron and MAWTS-1. They brought us up to speed on the rollout of the F-35B to IOC next year; MAWTS-1 provided us a perspective on the evolving tactics and training for that aircraft, as well as discussing with us the various roles, which MAWTS-1 performs for the Marine Corps air/ground team, both for Marine Corps Aviation and the MAGTF.

We had a chance to talk with the Operations Officer of MAWTS 1, Major Douglas Seich. Among other discussion points was an overview on MAWTS-1, its mission sets, its activities, its working approaches, and the way ahead.

The role of the squadron is to provide standardized advanced tactical training and certification of unit instructor qualifications that support Marine Aviation Training and Readiness. MAWTS-1 also provides assistance in the development and employment of aviation weapons and tactics.

The squadron works with new and older aviation assets and aims to train the trainers. It shapes ways to work in the new with the old and in supporting the infantry units in the MAGTF globally. It works with other services in shaping tactics and training approaches. Given its location it works closely with the USAF in Nellis, and with the F-35 will work closely with Luke and Nellis in shaping new tactics and training approaches as that aircraft enters the fleet.

The squadron runs two Weapons and Tactics Instruction (WTI) courses a year and various activities and exercises with Marines and others to shape the evolving future concepts of operations as well.

The core tasks identified in the briefing by Major Seich were as follows:

> 1) Conduct two Weapons and Tactics Instructor Courses (WTI) annually.
> 2) Support the Marine Air Ground Task Force Training Program's (MAGTFTP) Integrated Training Exercise (ITX) and Large Scale Exercise (LSE).
> 3) Conduct aviation training support (flight and academic) at the operating force's home station in order to provide advanced instructor certifications and 2000 - 6000 training and readiness support.
> 4) Manage all aviation tactics publications like the Air Naval Tactics, Techniques, and Procedures (ANTTP) manuals and the Assault Support Tactical Standard Operating Procedure (ASTACSOP) manual.
> 5) Serve as the syllabus sponsor for all Aviation and Tactical Air Control Party Training and Readiness manuals.
> 6) Provide representation at Aviation conferences like the Marine Air Board (MAB), Operational Advisory Group (OAG), and platform specific Transition Task Forces (TTF).
> 7) Maintain liaison with all branches of the U.S. Armed Forces, select foreign forces, and other military or technical organizations in order to ensure aviation tactics are current and appropriate.
> 8) Execute Tactical Development and Evaluation (TAC D&E) projects on behalf of Marine Aviation.

Credit: MAWTS-1

The size of MAWTS is around 200 total personnel with about half of those being officers. MAWTS obviously grows during the time the WTIs are conducted. For example, with regard to the first WTI this year there were 236 students involved, but with them an additional 92 aircraft and more than 4000 maintenance personnel and 2350 sorties generated.

The basic WTI course design focuses upon "training the trainer." And covers a number of key elements: offensive air support, control of aircraft and missiles, assault support, aerial reconnaissance, anti-air warfare, electronic warfare and aviation ground support.

Major Seich provided a look at the creation of new units and capabilities within MAWTS over the past 10 years. 2004 saw the establishment of an Air Officer Department working the role of the ground-to air integration efforts. He highlighted that the availability of joint training money was crucial to the functioning of MAWTS and that in 2005 they established a Joint National Training Capability and a Tactical Risk Management capability as well.

He noted: *The services have recognized the need to work more effectively together. The service chiefs have created the Joint Training Capability Fund (JTCF) to support such efforts. The Fund provides money and assets to work*

MAWTS-1

together. For example, the Marine Corps does not have enough fourth-generation aircraft adversaries for our exercises. We bring down USN and USAF platforms to play that role. To pay for that funding comes from the JTCF.

Majors Greenwood and Seich Outside of the MAWTS-1 building after our interview.

In 2011, the squadron created a Re-Black program which is designed to bring back MAWTS graduates for a three-day event to provide them updates with new tactics and training approaches since they have gone through MAWTS. And in 2012 MAWTS started focusing on training and tactics for long-range tiltrotor assault capabilities with the Infantry Battalion Training or TALONEX.

And last year two new divisions were set up, the F-35B division and the Spectrum Warfare Division. The Spectrum Warfare Division is a Division within C3 which focuses on: cyber domain operations, EW operations and MAGTF integration. And an area of interest is clearly working on the Anti-Access/Area Denial challenge. And obviously standing up the F-35B and Spectrum Warfare Division provides the opportunity for them to be mutually supportive and interactive.

Clearly, MAWTS is a living organism, which builds interactively upon real world combat experience.

Question: How does MAWTS integrate lessons learned from combat?

Major Seich: *We do that in a number of ways.*

The students coming through the course have just come from recent combat experience. And we send MAWTS instructors to combat theaters for about a month at various points and they bring back lessons learned.

We are able to look at what we do not know and what we need to know based on evolving combat situations. What is changing that we do not know about?

By and large, MAWTS uses the ranges in the Southwest of the United States to perform its exercises and training.

And the interactivity with the USAF, USN and US Army is important. With regard to the U.S. Army working with the Patriot has been significant, and the USAF has operated various assets including F-22s, F-15s, and F-16s and the Navy with F-18s are all important for the WTIs.

Interestingly, the F-22s from Nellis have participated and have functioned as part of an assault force.

According to Major Seich:

We have operated with F-22s to augment the Blue Forces working with us to perform certain functions.

Question: Have you focused on F-22 and 4th generation aircraft integration by working with F-22s?

Major Seich: *We have for a number of years worked on this with the USAF.*

For example, the F-18 and the C3 shops put on an air tactics course twice a year.

This focuses largely on air- to air. It is a four-week course and in 2010 we did an integration approach with the F-22, which was quite successful.

Question: With the work you have done, could an F-22 package work with an ARG-MEU in supporting a USN-USMC assault package?

Major Seich: *Absolutely. Foreign participation can occur as well in the WTI courses, and the United Kingdom, Australia, Canada and Israel have all sent participants in various WTI events.*

Reflecting on the Tiltrotor Enabled Assault Force: The Perspective of a MAWTS-1 Osprey Instructor

July 30, 2014

During our time at MAWTS-1, we had a chance to talk with Captain Justin "Lumbergh" Sing who represents the new generation of Osprey operators who have not transitioned from other platforms. He noted:

I have not flown any other fleet aircraft. I went through the flight school syllabus and straight to the MV-22 FRS.

Captain Sing had just joined MAWTS and had been there only three days when we interview him.

He had two tours at sea with the 26th MEU as part of VMM-266(REIN). The 26th MEU was involved in the Odyssey Dawn Operation, but Captain Sing was part of the split Osprey force and was serving in Afghanistan during that operation. Sing served under Col. Romin Dasmalchi for his first tour and Lt. Col. Christopher Boniface for the second.

During his time in Afghanistan, the Marines were expanding the operational envelope for the Osprey. As he noted:

We started utilizing V22 aircraft for the named operations in a new area previously unoccupied by US forces while I was there.

Question: Day and night missions?

Captain Sing: *Both. The first named mission I ever flew was a night insert.*[1]

The V22 community initially had issues with the dust cloud associated with brownout landings imposing an additional component to the "fog of war" encountered by the grunts during inserts.

It got worked out through Tactics, Techniques and Procedures (TTP) development and a good working relationship with the infantry.

He described one mission in Afghanistan in which the Osprey landed Marines and then quickly came back to move them out of harm's way. The quick turn-around capability of the Osprey is an important capability for the "devil dogs" coming out of the back of an Osprey.

Captain Singh noted: *Two Ospreys inserted troops to a particular landing zone, one on either side of a tree line. We departed and repositioned to a laager point about 15NM away. Fairly soon after, we were called back to move the Marines out of a suspected IED infested area. They could not safely cross the tree-lined ditch at night.*

The next day we found out that the Landing Zone (LZ) where we had conducted the insert had IEDs in it. We just happened to not land on any. That was our first operation after our unit had just arrived in Afghanistan.

Captain Sing highlighted the quick turnaround time, which the Osprey was able to provide to the troops on the ground.

From the time they called for immediate re-embark when we were on deck at the laager point, to the time they were repositioned, which included us landing, them loading, and us hopping the tree line and landing again— it was probably less than 15 minutes.

Captain Sing highlighted the impact of speed in an emergency medical situation as well.

We were onboard the ship and had a sailor with a gallbladder issue. It was about to rupture, and they needed to get him to a medical facility.

We were just north of Somalia in the Horn of Africa, and the closest medical care facility was in Mombasa down in Kenya. This happened while a party was being held on the flight deck, with no flight ops schedule that day. We needed to get this guy to medical care.

The deck crew cleared the front half of the boat and pulled the V22 out on spot within 45 minutes, and we were in the air 45 minutes later. We had to tank on the way, but we had him on deck in Mombasa, Kenya roughly 1,100NM away within 4+30 hours after takeoff.

When asked how the Osprey had advantages over rotorcraft in approaching Landing Zones (LZs), the Captain highlighted the advantage of a lower audible signature.

We can maintain an audible standoff for a little bit longer by staying in airplane mode up at altitude and only descending when approaching the objective area. It really reduces the enemy's ability to know we're coming.

The aspect of range was highlighted by a self-deployment discussed by Captain Sing.

When we were complete with required operations, we self-deployed from Afghanistan back to Greece. It was the longest flight a V22 had done at the time.

It was longer than the flight across the Atlantic to go to Farnborough, which had been the longest flight before.

The deployment was necessary as part of the force build up for Odyssey Dawn.

Captain Sing after our interview.

As Captain Sing underscored: *We had four planes conducting alert for Odyssey Dawn when the TRAP mission in Libya was executed. In order to reconstitute the unit, we flew three V22s and two C-130s in one day, a 16-hour flight total time, from Afghanistan to Souda Bay.*

The C-130s turned around the next day, flew back to Afghanistan, and then the following day conducted the flight again with the remaining three V22s. All six V22s, flew a one-shot from Afghanistan to Souda Bay over a three-day period. The only real limiting factor in the time period for execution was the external tanker availability.

Captain Sing highlighted that the Osprey was both a very easy plane to fly but an unforgiving one as well if it was not properly respected.

It's an easy plane to fly, it really is. In VTOL you're dealing with a rotor head with a vectored thrust component, and there's a lot of aerodynamics at play. The vectored thrust is what makes it unique and allows for an increased range of capabilities; but where we really see the gains are in the ability to fly like an airplane during the en route portion of any mission. It's a great aircraft with a lot of capability.

The Captain was asked about his focus at MAWTS-1 on the next steps for the Osprey.

I think digital interoperability is the next step for us. Also, the over-the-horizon capability, figuring out how we're going to provide ourselves some eyes on the objective area before we get there, as we're going to be the ones with the range capability to conduct those over the horizon inserts.

The Captain provided a sense of how he saw the path towards this end state.

We're working on flying with a bigger communications package that allows us to have digital interoperability in communications with most wave forms. This would increase our capacity to inform the ground force commander in real time on the way to the objective area and increase his situational awareness to help ensure preconditions for insert have been met prior to placing boots on the deck.

The ability to pull information from multiple sources in real time would be a significant enhancement to capability.

Visiting the F-35 Squadron at Yuma Air Station: The Executive Officer of VMFA-121 Provides an Update

August 6, 2014

We had a chance to discuss the work of VMFA-121 and to get an update with the Executive Officer of the Green Knights, Major Gregory Summa during our visit. The aviators and maintainers of this storied squadron are working to bring to the first F-35B Squadron into service next year.

Historically it is interesting to note that VMFA-121 was activated in June 1941 and began flying air ground combat missions in August 1942, with the "Cactus Air Force" on Guadalcanal. The Green Knights made Marine aviation history with 14 aces, including the legendary Joe Foss CMH so IOC means just that, ready for combat.

Question: How would you describe the current role of the squadron?

Major Summa: *The Marines focus on a process of giving the airplane to the operators and let the operators figure out how best to operate and then use the aircraft.*

Our leadership has prepared the way for the coming of the F-35 to the USMC and has worked hard to ensure that the infrastructure is in place to allow us to train and use the aircraft. For example, when LtGen Trautman was Deputy Commandant of Aviation he focused on preparing Yuma to be the home for the first F-35 squadron.

Clearly, being here with MAWTS-1 gives us a good advantage to get a good start on operating, training, and shaping the tactics of the new aircraft for the MAGTF.

After creating the infrastructure, the next step was to get the airplane in the hands of Marines to work with the aircraft and to work with the aircraft within the limits of what it is cleared to do, because we do not have clearance for the full flight envelope we will have by the time the aircraft attains Initial Operational Capability.

Question: Putting the plane in the hands of the operators is a key part of developing the aircraft as well isn't it?

Major Summa: *It is. Every time we fly, we are learning something.*

While trained Test Pilots are operating instrumented aircraft on a detailed test plan, in Yuma you have operational pilots flying the jet everyday gaining data points that may not have been discovered by Developmental Tests.

By data points I do not mean safety of flight related items, I am referring to operational data points. More along the lines of how to optimize and use the multiple sensors to accomplish a task or execute a mission set.

Since we have such a good working relationship with the Developmental Test entities, the Joint Operational Test community, and the individuals from industry who are SMEs on the systems, we can get immediate feedback when questions arise and then promulgate that back out to the community.

For example, last week we spent several hours in the vault with pilot training officers and with pilots who have been either MAWTS or Top Gun graduates or instructors. We compared our operational experience with what has been developed so far with regard to our joint tactics manual which was written more than year ago, based on expectations developed from flying in the simulator.

Now we are seeing things in the operational airplane. So how do we change? How do we improve, update and morph the manual to where we see the plane operationally performing?

Question: How do you externalize your learning outside of the squadron?

Major Summa: *One way is working with the USAF at the 422 Test and Evaluation squadron at Nellis. We tend to be busy here, so we send operators from the training department or former patch wearers (MAWTS-1 and TOPGUN) to work with SMEs from the Navy and USAF at conferences or simulator events.*

The young senior company grade who are coming off of a tour with a Hornet or a Harrier and now wearing a Green Knights patch go into the room with the aviators at Nellis with F-16 and F-15 pilots and work through the process.

In effect, an F-35 enterprise is emerging built around a group of individuals in the profession of arms who want to make this airplane as lethal as possible. People come in from different backgrounds – Raptor, Eagle, Viper, Hornet or Harrier – and are focusing on the common airplane and ways to make it work more effectively in a tactical setting.

And talking to the experience of a common plane is a crucial piece of the effort. When an F-35 pilot sits down regardless of what service he is in, he's talking with an individual from another service on the same data point.

Let me explain what I mean. If I sat down as an F-18 pilot, and I wanted to talk about AMRAAM performance, I was talking about it relative to how it integrated with an F-18. The F-18 is a Boeing product, a McDonald Douglas product, totally different than F-16, which is a Lockheed product.

When I talk AMRAAM with an F-35 pilot from the Air Force, maybe one of the squadrons that is based at Luke Air Force Base. I am talking about the same exact radar; I'm talking about the same exact software — everything's the same. If we differ in training, it doesn't have to do with hardware, it doesn't have to do with software; it has to do with service approaches or carry-over from previous doctrinal employment.

When an F-35A pilot talks with an F-35B pilot and they discuss what they would to see with the evolution of the aircraft they are discussing essentially the same airplane and its evolution. It is two operators of the same airplane focused on what they want to see evolve even though they are in different services. And the commonality point is really lost in the broader discussion of the F-35.

And when it comes to strategic impact it is the commonality associated with logistics, which will have a really significant operational impact. The interoperability at the supply level, the logistics level, the procurement level or the mainte-

nance training level is a key foundation for joint and coalition airpower going forward leveraging the F-35.

Question: Let us focus on the squadron and its composition and work schedule, so to speak. What is the current situation?

Major Summa: *We have 16 airplanes in the squadron. We have 15 pilots who have gone through VMFAT-501 at Eglin. Nine of those pilots have gone through S/TOVL training and are qualified completely to operate the plane that way. The others will complete the syllabus shortly.*

Question: Are these primarily Hornet or Harrier pilots?

Major Summa: *We only have F-35 pilots. Our flight temp is Tuesday through Friday. We have the only organic maintenance department in DOD. When I say organic, I mean that we do not have contractors fixing our airplanes, we have Marines fixing our airplanes. We have the normal technical representative support from contractors as one would expect with an organic squadron. We are 260 strong in the squadron and we run two shifts, six five-days and six five-nights a week. Our pilots fly around 15 hours per month.*

Major Summa after our interview at MAWTS-1.

Question: When you fly the plane how do you balance the air-to-air and close air support missions?

Major Summa: *That is a good question. The plane and its combat systems and the way the cockpit is designed allows the pilot to handle the missions in a very effective an integrated manner.*

To be able to do CAS, you have to make certain that you can suppress threats that would make it prohibited. With this plane, you can affect the environment to make CAS more readily available and more quickly.

Question: The F-35 is a multi-tasking aircraft and as such how do you approach doing air-to-air and air-to-ground missions?

Major Summa: *You can flip between the two without ever forgetting where you were on the last one.*

And let me explain that a little bit better. In the F-18, when we were going to air-to-ground mode specifically on the strike, and we are using the radar, and if we want to use the targeting pod, we would get to a certain point in time in the mission, where we have to use some sort of a planning tool. The pilot would have to sort out when he would be able to go all heads down to try to find the target and employ on the target.

There is an operational difference as well. I need to have a certain amount of distance between me and a threat so that when I come heads back up and start looking for possibly an air breathing threat or a surface-to-air missile, would need to suspend the task of employing that piece of ordinance or that weapon for the CAS mission.

This airplane's different because with the data being fused, I'm not using multiple different displays with each.

The main difference that I see between federated and fused systems is in the F-18, not only was it all in different displays, but each sensor had its own uncertainty volumes and algorithms associated with it. It was up to me as an aviator knowing the capabilities and limitations in my system to decipher and draw the line between the mission sets.

In the F-35, the fusion engine does a lot of that in the background, while simultaneously, I can be executing an air-to-air mission or an air-to-ground mission, and have an air-to-air track file up, or multiple air-to-air track files, and determine how to flip missions.

Because the fidelity of the data is there right now, which allows me to deter-

mine if I need to go back into an air-to-air mindset because I have to deal with this right now as opposed to continuing the CAS mission.

And I have a much broader set of integrated tool sets to draw upon. For example, if I need an electronic warfare tool set, with the F-18 I have to call in a separate aircraft to provide for that capability. With the F-35 I have organic EW capability. The EW capability works well in the aircraft. From the time it is recognized that such a capability is need to the time that it is used requires a push of a button. It does not require that a supporting asset be deployed.

Question: Obviously your pilots need to be trained to combine the air-to-air and CAS capabilities and to use the new organic tools sets as well?

Major Summa: *It does. Now we're going to have a pilot that's versed in doing CAS, if he needs to use the electromagnetic spectrum or exploit it to accomplish his mission, he'll be educated and have the equipment to do so. If he needs to use it in the air-to-air arena to exploit it, to accomplish his mission, he'll have the training and the equipment needed to use it as well.*

In the current situation, I would deploy a Prowler to work with my legacy fighters. The Prowler would have to be sortied and would operate only for a period of time and in a specific operational area. With the low observability of the F-35 combined with the organic EW capability of the aircraft, the aircraft expands my capabilities for both air-to-air and CAS.

Working the Tactics and Training of F-35Bs with VMFA-121: The Perspective of Maj Roger "HASMAT" Greenwood

August 10, 2014

Major Greenwood is one of the two MAWTS-1 officers involved with the F-35 and standing up the initial division within MAWTS to develop tactics and implement training for the new platform in order to integrate it into the MAGTF. The other is Major Noble.

Major Greenwood: *We are really starting to stand up the capability of the F-35 and working on its integration. We have started flying the F-35 in WTI events here at MAWTS, one of which was an event called AAW2, which is a big defensive counter-air event that we run during WTI, and has been the biggest event that they've flown in to-date.*

We had some fairly good success as well. It was pretty eye-opening, I think, for a lot of people to see the capabilities that the aircraft brings, even in a 2A configuration. We were able to do some pretty impressive things in this event, which highlighted things to come as well, notably with the radar.

AAW 2 is an air-to-air event defending high value ground based MAGTF assets from a threat strike. The F-35s integrated with F-18s and a notional Patriot battery against adversaries, which included, F-18s, F-5s, AV-8s, EA-6Bs, B-1Bs.

The fidelity of the radar is amazing. That sensor is obviously well beyond anything that we have in our F-18s. We can see things that the Hornets weren't able to see, and then right now, passing information via voice only to the F-18s in two-day 2A aircraft.

We will have the data link capability in the next block of software which is coming shortly. And more generally, as the aircraft enters service it will become a key factor in keying up other assets, such as the F-18s to provide additional firepower identified by the F-35 sensors.

Question: And the DAS 360-degree sensor system along with the radar all by themselves presents new capabilities for you as well?

Major Greenwood: *They do in terms of our ability to see things we could not see before and they give us significant advantages in the battlespace.*

Major Greenwood after our interview with him at MAWTS-1.

Question: You already have a training and tactics manual, how is that progressing?

Major Greenwood: *We do have a basic manual, but the approach is in some ways along an Air Force Model whereby we develop the qualifications for an instructor pilot as the basic focus.*

Question: What is your relationship with the other services in rolling out the F-35?

Major Greenwood: *We have a habitual relationship with the USAF 422 Squadron which is their test and evaluation squadron at Nellis as well as with their 31st Squadron at Edwards AFB which is a test and evaluation squadron as well. And through them we see other units as well who are engaged in preparing for the integration of the airplane into their services.*

The Navy interface is pretty small as we deal primarily with one officer from Top Gun. They are just getting their feet wet. They have been involved in the process throughout.

Question: There is a unique role for the USMC as the first service to operate the aircraft, but obviously working with the USAF is important as well for the USMC. Could you describe this relationship?

Major Greenwood: *It is an important one. Because we are going first, there is obviously a keen interest in what we are doing here at Yuma.*

But we are working closely with the other services as they prepare to operate the aircraft. The USAF is especially important in this regard.

But obviously, the USMC is in a unique position here. And as we prepare for IOC, we are shifting from a requirements role to a training role with regard to the aircraft.

The working relationship with VMFA-121 is obviously central. They're starting their IOC training through the T&R progression next month. And we will obviously be involved in that process, and with the ultimate goal being to get them to IOC, and then eventually to have students coming through WTI.

Question: What weapons will the Marines be operating with the IOC aircraft?

Major Greenwood: *In a Block 2 aircraft, we will be able to carry two AIM-120s and either two GBU-32 JDAMs or two GBU-12 laser guided bombs internally. External load outs will start with the block 3-F configuration. All of the combat systems will be functional and obviously will evolve over time in the software upgrades.*

Question: It must be exciting for you working on the

initial operational tactics and training with a new generation aircraft?

Major Greenwood: *It's always exciting, I think, to be at the tip of the spear, if you will, which is kind of where the Marine Corps likes to operate anyway. It is a unique opportunity and a very good opportunity for us.*

A VFMA-121 Maintainer Provides an Update on the Maintenance System for the F-35

August 8, 2014

As the first squadron to maintain an F-35B, the VFMA-121 maintainers are clearly key players in shaping the operational reality of the F-35 and its future. The often bandied about term "concurrency," which is usually used to criticize the F-35 program actually underscores a strength: the operators – pilots and maintainers – are in a position to shape the roll out and evolution of the F-35 enterprise.

When visiting Yuma Marine Corps Air Station in July 2014, we were able to discuss the next evolution of the maintenance regime, namely a squadron maintained by its own organic assets.

VMFA-121 is the first F-35 squadron and the first with organic maintenance. A squadron with organic maintenance simply means that the Marines are manning the maintenance squadron with inputs from technical representatives, but because it is the first operational squadron obviously the Marines need to prepare for overseas deployment and to prepare to support the aircraft in forward positions.

Notably, the squadron has already deployed for movement to the United Kingdom for the Royal Tattoo and Farnbourgh Air Shows but was stopped at Pax River while DOD made its decisions on the go ahead with F-35 fleet engines, a process that concluded favorably but too late to permit the squadron jets to fly across the Atlantic.

They had to fly back across the United States to Yuma on the day we were visiting the squadron. And the flight to England was viewed as part of the overall progress to the IOC of the aircraft next Summer.

MAWTS-1

Staff Sargent Jason Lunion after our interview with him at MAWTS-1.

As part of that progress, the maintainers from the squadron accompanied the jets and were prepared to support the plane fully in operation. In the work up for RIAT/Farnborough, VMFA-121 conducted the first ever engine change away from home station at Pax River. Installation went quicker/smoother than was predicted and helped VMFA-121 move closer towards having a combat/expeditionary IOC deployment capability in 2015.

We had a chance to discuss the progress with a powertrain maintainer on the F-35 working at VMFA-121. Staff Sargent Jason Lunion has been a maintainer since 1999 and his first squadron CO (for VMFA-223) was LtGen Davis who is now the Deputy Commandant of Aviation.

Staff Sargent Lunion most immediately comes from working on engines with the Harrier but has wide range of experience, as one would expect for members of the first squadron with organic maintenance for the USMC in supporting the F-35.

The F-35 is the first low observable aircraft to be operated by the maritime services, and requires some changes in how the maintainers support the aircraft, and notably at sea.

The discussion with the Staff Sargent highlighted that the low observable qualities of the aircraft created some specific challenges, and one of those, which he mentioned, was working on the panels. The panels on the aircraft provide easy access for a number of maintenance functions, but as he described it one change is the impact on the T-handles, which open the panels.

He noted: *The panel is opened numerous times a day and we are wearing down T-handles that provide access to the panel and wearing down the fasteners themselves.*

He was asked about the general shift from legacy to LO maintenance and highlighted that the Marines have not operated an LO aircraft before so there is a learning curve.

He underscored: *There is a drastic increase in awareness when you are working around the aircraft.*

A key aspect of the aircraft is the use of computer aided maintenance and sensor-informed systems.

The Staff Sargent focused on how the sensor-enabled aircraft was also a work in progress much has one has seen with new commercial aircraft which rely heavily on sensors to provide data about performance and maintenance demands. He cautioned: *When a sensor indicates a problem, is it the sensor or is a real problem? Because of all these sensors, and all these little gadgets on the motor that are supposed to eventually take this to an on-condition inspection basis, until their maturity's reached, we're going to continue to have a lot of fine tuning to do.*

He noted that compared to the Harrier working on the F-35 engine was much easier. *With regard to the Harrier, you have to remove the wing and then crane the engine out. That is clearly not very maintainer friendly, but the F-18 is a different case where removal of the engine is straightforward.*

When asked about his overall experience, he emphasized that some aspects were welcome additions, and others were a work in progress.

We were told it was designed to be maintainer friendly, and obviously, we're finding as with any platform, you're going to find certain things that are maintenance friendly, and there's certain things that are not. But we haven't really found anything that's extremely difficult working on this aircraft.

I think the biggest challenges to date are due to the immaturity of the technical data that we used to fix the plane.

He highlighted that the lift fan was turning into a good maintenance experience. *The lift fan isn't as hard as anybody was originally thinking when we started. The tolerances in the airframe for pulling a motor is so tight in a Harrier that you, at times, have to literally shake the motor free of some of the stuff that's in the way on the airframe. With the F-35, this is not a problem.*

His major complaint was that the unit wanted to ensure that they were able to get into a position whereby the maintainers could be able to work with the whole aircraft and maintain it as a single entity.

I think the single biggest difference going from a Harrier community to coming here is not the maturity of the aircraft because that was expected.

It is the lack of technical publications that really are impediments.

We continue to work towards a comprehensive and complete technical pubs library to realize/maximize the efficiencies the F-35 brings.

4

The 2016 Visit

We visited MAWTS-1 in December 2016. With the forthcoming deployment of the Green Knights to Japan our visit was really dominated by that upcoming deployment.

A January 2017 press release regarding the deployment to Japan naturally highlights why the Green Knights were the focus of attention.

MARINE CORPS AIR STATION MIRAMAR, California (January 10, 2017)

Marine Fighter Attack Squadron (VMFA) 121, an F-35B squadron with 3rd Marine Aircraft Wing, departed Marine Corps Air Station Yuma, Arizona, transferring to Marine Corps Air Station Iwakuni, Japan, Jan. 9, 2017.

The first location to receive the Marine Corps' F-35B, as part of its worldwide deployment capability, is Iwakuni, Japan.

In November 2012, the Marine Corps announced that after a century of Marine Corps aviation, 3rd Marine Aircraft Wing would introduce its first F-35B Lightning II squadron. The F-35B was developed to replace the Marine Corps' F/A-18 Hornet, AV-8B Harrier and EA-6B Prowler. The Short Take-off Vertical Landing (STOVL) aircraft is a true force multiplier. The unique combination of stealth, cutting-edge radar and sensor technology, and electronic warfare systems bring all of the access and lethality capabilities of a fifth-generation fighter, a modern bomber, and an adverse-weather, all-threat environment air support platform.

An F-35B Lightning II, assigned to the "Green Knights" of Marine Fighter Attack Squadron (VMFA) 121, lands at Misawa Air Base. VMFA-121 is the first operational F-35 squadron in the U.S. military. January 9, 2017. Credit: Naval Air Facility, Misawa.

Nov. 20, 2012, VMFA (All Weather)-121, formerly a 3rd MAW F/A-18 Hornet squadron, was re-designated as the Corps' first operational F-35 squadron, VMFA-121. The Commandant of the Marine Corps publicly declared VMFA-121 initial operating capability (IOC) on July 31, 2015, following a five-day operational readiness inspection (ORI). Since IOC, the squadron has continued to fly sorties and employ ordnance as part of their normal training cycle.

In December 2015, VMFA-121 employed its F-35Bs in support of Exercise Steel Knight. The exercise is a combined-arms live-fire exercise which integrates capabilities of air and ground combat elements to complete a wide range of military operations in an austere environment to prepare the 1st Marine Division for deployment as the ground combat element of a Marine Air-Ground Task Force (MAGTF). The F-35B preformed exceedingly well during the exercise.

In October 2016, a contingent of Marine Corps F-35B's, pilots and maintainers participated in Developmental Test III and the Lightning Carrier Proof of Concept Demonstration aboard the USS America (LHA-6). The final test period ensured the plane could operate in the most extreme at-sea conditions, with a range of weapons loadouts and with the newest software variant.

Data and lessons learned laid the groundwork for developing the concepts of operations for F-35B deployments aboard U.S. Navy amphibious carriers, the first two of which will take place in 2018.

The transition of VMFA-121 from MCAS Yuma to MCAS Iwakuni marks a significant milestone in the F-35B program as the Marine Corps continues to lead the way in the advancement of stealth fighter attack aircraft.[1]

The Way Ahead for USMC Con-ops: The Perspective of Col Wellons, CO of MAWTS-1

December 30, 2016

During our most recent visit to USMC Air Station Yuma, we had a chance to meet with the head of MAWTS-1 and discuss the way ahead for USMC concepts of operations as seen from this key tactical innovation center.

Question: When we were last here in 2014, MAWTS-1 did not yet have its own F-35s.

Now you do.

How are you working its integration with the MAGTF?

Col Wellons: *The great thing about MAWTS-1 is we run the Weapons Training Instructor course at Yuma twice a year, and as a former CO of MAWTS put it to me, WTI is where the USMC comes together every year to train for war.*

We are able to do the high-end training in terms of aviation support to the MAGTF. The F-35 is integrated into every mission that we do, whether it is close air support, helicopter escort, or, at the high end, air interdiction operations against a high-end threat including integrated air defense as well.

When we come back from a typical WTI mission exercise, and we debrief it with the helo and fixed wing guys and the C2 guys and the ground combat guys, more often than not it is the F-35 which is identified as the critical enabler to mission success.

It is the situational awareness we gain from that platform, certainly when dealing with a higher end threat like dealing with air defense, that provides us with capabilities we have in no other platform.

I am pleased with where we are with the airplane right now. We have declared IOC and we are getting to deploy it to Japan.

Question: How does the integration of the F-35 into your operations change how you think about those operations?

Col Wellons: *A lot of that can be quickly classified but let me give you an example, which does not fall into that category.*

Historically, when we could come off of L class ship with Mv-22s, CH-53s, Cobras and Harriers facing a serious AAA or MANPADS threat we would avoid that objective area.

Now we do not need to do so. It changes the entire concept of close air support.

In Afghanistan and Iraq, we have not had prohibitive interference in our air operations. With double digit SAMS as part of threat areas we are likely to go, the F-35 allows us to operate in such areas. Without the presence of the F-35, it would be a mission that we wouldn't be capable of executing.

The SA of the airplane is a game changer for us. Rather than getting input from the Senior Watch Officer on the ground with regard to our broader combat SA, we now have that in our F-35. This allows us to share SA from the pilot flying the airplane and interacting with their sensors. They can share that information, that situational awareness, with everybody from other airborne platforms to the ground force commander in ways that are going to increase our ops tempo and allow us to do things that historically we wouldn't have been able to do.

The ability of the F35 to be able to recognize and identify the types of prohibitive threats that would prevent us from putting in assault support platforms and ground forces is crucial to the way ahead. The F-35 can not only identify those threats, but also kill them. And that is now and not some future iteration.

Question: You are innovating as well with the F-35 as you integrate with your forces. Can you describe an example of such innovation?

Col Wellons: *Absolutely.*

One example has been something we did in the last WTI class, namely hot loading of the F-35 as we have done with the F-18 and the Harriers in the past.

We worked with NAVAIR and with China Lake and Pax River and came up with a set of procedures that we can use to do the hot load of an F-35. We did it successfully at this last WTI class, and it significantly shortens the turn time between sorties.

When you think about us operating in some places around the world that we do, the number of additional sorties we can generate as a result of being able to do that, and the reduction in the vulnerability that we have in terms of the turn-around is crucial.

Also, whenever you shut an airplane down, whether it's a fifth-gen airplane or a legacy airplane, it has a greater tendency to break. We did GBU-12 last class, we'll be doing GBU-32 and AIM-120 this upcoming class.

Question: Obviously, you are working with the USAF and the U.S. Navy on reshaping air operations affecting the MAGTF. Can you give us a sense of that dynamic?

Col Wellons: *For the USAF, the capabilities of the airplane in terms of the sensors that we have, the weapons that we have, the way that we're employing this airplane, they're remarkably similar.*

We are in lockstep with Nellis, with the weapons school, with the 53rd Tests and Evaluation Group in terms of how we're doing operational tasks, and we are very closely aligned with them in terms of how we employ the airplane, how we support the airplane.

We do quite a bit of work with Fallon. They are on a different timeline from the Air Force. They're a couple of years behind in terms of where they are, but I anticipate that we'll have similar collaboration with the Navy as they begin to lean forward into the F35 in the next couple of years.

Shaping a 21st Century Assault Force from the Sea: The Perspective from VMX-1

December 29, 2016

Col Rowell is the first Commanding Officer of VMX-1: Marine Corps Operational Test and Evaluation Squadron 1. VMX-1 includes the operational test & evaluation (OT&E) and science & technology (S&T) activities that have supported Marine Aviation from HMX-1, VX-9, MACCS-X and MAWTS-1.

One of its predecessors was VMX-22, which was established in 2003 for the express purpose of introducing the Osprey and shaped its evolving con-ops. More than a decade later the Marines of VMX-1 are now helping to integrate the F-35B into Marine Air Ground Task Force (MAGTF) and are preparing for the next new Marine Aviation asset, the CH-53K.

We started the interview with Col Rowell by recalling the original VMX-22 in this manner. He commented: *When we were setting up the office, a Marine came in and said we had some old gear we needed to dispose of, including an older flight helmet. I turned the helmet around and the name on it was Walters. It now occupies the top shelf in my office.*

About 2009, the OT&E missions of HMX-1 were ported over to VMX-22 to work through innovations with the CH-53E. The same had been done with the attack and light lift/utility helicopters years earlier with VX-9. All of those missions, along with the F-35B and CH-53K efforts, have taken root in VMX-1 as well.

The unit is one which now has its foot firmly planted into the future while simultaneously shaping today's fight. VMX-1's F-35Bs are at Edwards AFB as part of the Joint Operational Test Team which is working with their developmental test counterparts to evaluate and integrate the ongoing upgrades of the aircraft.

The VMX-1 F-35Bs will come to Yuma in 2018 and will be the center of excellence for global F-35Bs as well after the Block 3F software is complete. VMX-1 will continue to shape the demand side for the F-35B community with regard to upgrades as well.

We asked about how integrated the British have been with Rowell and his Marines.

Col Rowell noted that there is very close integration. *It is crucial. We carrier qualified a Royal Navy pilot onboard the USS America in USMC airplanes. We are exchangeable. There is no light between the Brits and the Marines. On the USS America, you had UK maintainers, and you had observers from HMS Queen Elizabeth on board the USS America as well. It is very important for the community to remain focused on commonality. There is widespread recognition of this requirement.*

The Marines are a key stakeholder in this process with the services and the allies. We are well tied into the community to shape commonality for upgrades and shaping the way ahead. This applies in strategic terms to shape integrated airpower from the UK to Norway to Denmark to the Netherlands and operating off of U.S. and UK seabases.

The interoperability between the USMC and the UK is a key thread in that effort with our ability to operate off each other's ships. It is like flying with someone else nationally but part of your own squadron.

We then asked about the status of how the maintainability aboard the USS America during recent tests.

We took an aircraft and pulled the engine, drive shaft and lift fan — then reinstalled and flew it off of the ship in sea state three. We validated many of the toughest maintenance tasks at sea with that maintenance evolution, and that jet was one of the first planes off of the boat during the Lightning Carrier demonstration.

The two Yuma squadrons plus VMX-1 were working the maintenance and almost all of the maintainers had never been to sea as well. Availability and maintainability were good. We did not lose any flying time due to maintainability. Very unusual for an aircraft at this stage of the game.

The test community is shifting its focus on airframe testing to the software upgradeability dynamic. We are internalizing that.

The biggest item I saw was the growing realization of what a software defined, and upgradeable plane is all about. Many of your hardware dynamics are also about software. For example, with regard to the fuel pump, what it does and how it performs is software driven. You have to tweak the software a bit and you can get the fuel pump do what you want to do with it.

We then discussed the coming of the CH-53K to the USMC and the role of VMX-1 in that process.

Not only is the lift much greater and the maintainability significantly better, but the aircraft will play into the enhanced situational awareness (SA) with the F-35, along with the speed and range of the MV-22 as an assault asset. The pilot flying the F-35 will shape much greater SA to the MV-22 and the CH-53K as they inform and support the overall assault force. In effect, this is the flying infrastructure for the future MAGTF. We will continue to refine the tactics, techniques and procedures (TPPs) as the force matures as well.

Working the MV-22 With F-35 Integration: Shaping Future TRAP Missions in a Dangerous World

December 30, 2016

In addition to the interview which we had with the CO of VMX-1, Col Rowell, we had a chance to talk with LtCol Nelson, the XO of MAWTS-1 and Major Duke.

LtCol. "Cowboy" Nelson was on the deployment under the command of Lt. Col. Bianco when we conducted an interview with the first squadron of MV-22s which deployed to Afghanistan in early 2010.

In that interview conducted by telephone when the squadron was in Afghanistan, LtCol Bianco highlighted several key contributions of the aircraft to the fight. The most compelling point underscored by the squadron commander is how, in effect, the Osprey has inverted infrastructure and platform. Normally, the infrastructure shapes what the platform can do. Indeed, a rotorcraft or a fixed wing aircraft can operate under specific circumstances.

With the range and speed of the Osprey aircraft, the plane shapes an overarching infrastructure allowing the ground forces to range over all of Afghanistan, or to be supported where there are no airfields, or where distributed forces need support.

The envelopment role of the Osprey is evident in Afghanistan as well, whereby the Osprey can provide the other end of the operational blow for the ground or rotorcraft in hot pursuit of Taliban.

The Osprey can move seamlessly in front of rotorcraft and land forces, allowing the pursuit of different lines of attack. The envelopment role was not the focus of the interview because of security considerations, but anecdotal evidence suggests such an emerging role.[2]

The progress of the Osprey since then in terms of its performance and impact on the evolution of USMC concepts of operations has been significant. The F-35B coming into the force is having a similar impact but is building upon the prior experience, which the Marines have had with the Osprey.

Given "Cowboys" long experience with the Osprey and its

maturation, he brings the experience of Marines shaping a way ahead with revolutionary technologies associated with the Osprey to the new task, namely, the integration of the F-35 into the USMC.

He commented: *Part of the mission at MAWTS-1 is to familiarize the students with the new options associated with the F-35 and that requires a mind shift for Marines in working through how to best leverage the aircraft.*

The digital interoperability initiative being conducted by the USMC is a key part of shaping the situational awareness thread for the insertion of the assault force via the Osprey and the F-35. The F-35 as a key generator of SA to be distributed to the incoming assault force. "Cowboy" underscored: *The new generation is so technologically sophisticated that they will thrive in the evolving digital environment of which the F-35 is a key element.*

A key impact of integrating the MV-22 with the F-35 will clearly be with regard to the Tactical Recovery of Aircraft and Personnel (TRAP) mission.

The Osprey has already demonstrated a sea change with regard to how TRAP is done. This has already been demonstrated in combat in the Odyssey Dawn operation. With the USS Kearsarge off of (ironically enough) the shores of Tripoli, the Air Combat Element or ACE began to deliver unique resupply capabilities to the Kearsarge, which allowed the Harriers to triple their sortie generation rates.

By being able to fly directly to Sigonella rapidly and back, the Ospreys kept the Harriers in the air much longer than anticipated. The TRAP mission over Libya saw the Marines execute the mission at least 45 minutes faster than the next available platform and did so very rapidly after having received the go order.

Now with the F-35 replacing the Harrier and flying with the Osprey, the range of operational conditions into which the TRAP mission can now be flown is expanding significantly. Even into contested areas the F-35 can work with the Osprey to save lives and to extract pilots from harm's way.

This micro capability is reflective of what the USMC-USN team can do from the sea with an F-35 enabled force and be able to deliver the ground combat element via the Osprey backed by the F-35 as a significantly expanded close air support aircraft.

The dynamics of integration of the F-35 with the Osprey and changing concepts of operations provides political leaders with new strategic options for inserting and withdrawing force against a threat.

Ed Timperlake with LtCol Nelson and Major Duke after our interview.

We also discussed a key shift as the number of F-35s goes up, the role of the user groups will be enhanced in shaping the future evolution of the aircraft.

Major Duke noted that already at WTI courses NAVAIR engineers are coming to the courses and observing how the aircraft is being used by Marines in those courses. According to Duke: *They sent two engineers last Spring. This has happened in the past, so it is back to the future in effect as we shape the way ahead for the F-35.*

In short, the role of MAWTS-1 and its students will become key demand side drivers for how the software defined and upgradeable aircraft F-35 evolves.

The Green Knights On the Way to Japan: A Discussion with LtCol Bardo, CO, VMFA-121

January 6, 2017

We last visited VMF-121 prior it being declared IOC with the F-35B. During our most recent visit to MCAS Yuma we had a chance to visit both of the IOC F-35B squadrons in Marine Aircraft Group-13. We also visited with MAWTS-1 and VMX-1, who have just returned from DT-III testing onboard the USS America.

The first F-35B IOC squadron in the world, VMF-121, the Green Knights, are in the processing of transitioning to their deployment in Japan. Equipment and personnel are already on the way to Japan and the squadron will fly out this winter across the Northern Pacific to operate from Japan.

The deployment comes at a crucial time, given ongoing developments in the Pacific, and the opportunity to be combat operational with F-22s in Pacific Defense.

The F-35B will continue with this new generation of a V/STOL aircraft to work its flexibility with regard to ships and landing bases, which do not necessarily have to be regular airfields.

We had a chance to talk with LtCol Bardo, the CO of the squadron, who is taking the squadron to Japan but will soon thereafter transition from the squadron.

But Bardo has been with the squadron during its IOC and work up with the Marine Corps for its deployment to Japan. He and his squadron are performing key historical tasks as the cutting-edge operational F-35 squadron in the world.

This is an unusual situation for the Marines to find themselves in terms of combat air, but the flexibility of a combat information dominance aircraft fits right in with the evolving concepts of operations of the Marines. LtCol Bardo underscored the importance of Close Air Support for Marines and the role which the F-35 can play in significantly expanding the scope and nature of close air support.

MAWTS-1

Ed Timperlake with LtCol Bardo after our interview.

According to Bardo: *CAS is considered doctrinally a function which operates only in a permissive air environment.*

We can expand CAS to deal with a much wider range of situations than when we would simply operate in a permissive air environment.

And we can provide greater assurance to Marines as they deploy on the ground that we can deal with a much wider array of pop-up threats than we could do with legacy aircraft.

LtCol Bardo described the path to get to where the squadron was right now as it prepared for its Japanese deployment. The period since declaring IOC has been a busy and challenging one as the squadron pushed out the boundaries of the operational capabilities of the aircraft and worked with the Ground Combat Element (GCE) to integrate the airplane into the CAS role as well as working with the USAF on the air-to-air missions as well.

It has been a busy period for Bardo and his squadron but certainly historic as well. Throughout the squadron has found the core capabilities of the aircraft to be a solid foundation for shaping the way ahead.

VMFA-121 squadron F-35B jet pictured during our visit to MAWTS-1 in December 2016.

As LtCol Bardo described the F-35: *For the pilot, the ability to shift among missions without having to think sequentially about doing so is really a key strength of the aircraft.*

The airplane can think CAS and air-to-air at the same time and the pilot can then mix and match as the mission demands rather than having to think through the sequence of going from one mission set to the next.

In broad terms, LtCol Bardo described the progress of the squadron going from its time at 29 Palms working CAS, to working closely with MAWTS-1 on shaping the tactics for the use of the aircraft in support of the MAGTF, to its participation in Red Flag this summer as the F-35 component of the air operations being exercised at Red Flag.

In total, these experiences have been crucial in preparing the squadron for its deployment to Japan.

With regard to 29 Palms, the support to the ground combat element was the focus of attention in Steel Knight 2016, which included operating from Red Beach, an austere combat training facility where the presence of FOD or ground debris is a challenge.

At the exercise we could show Marines that the F-35 is a core asset for expanding the operational environment in which the MAGTF could operate and how we can support the GCE. We built trust in the infantry in what this revolutionary STOVL asset can bring to the force and to enhance their lethality and survivability as well."

MAWTS-1

With MAWTS-1, the squadron has worked closely on shaping the tactics and training for the new aircraft. The MAWTS-1 F-35 instructors have come from VMF-121, and the synergy has been crucial to shaping the way ahead for VMF-121 as it faces its deployment to Japan.

Then this summer, the squadron sent planes to Red Flag and flew in a US-only exercise with the full panoply of USN and USAF aircraft, excluding the F-15s. There the USMC flew its jets and were part of reshaping of air-to-air operations associated with the F-35.

A plaque in the MAWTS-1 bar remembering the participation of VMFA-211 in WTI 2-17.

LtCol Bardo noted that there were many F-16 National Guard pilots who were there, some of which had flown with the F-22 but had not flown with the F-35. They soon learned that you did not want to be an adversary but to leverage what the F-35 brought the fight.

As they prepared for the deployment to Japan the CO reflected on his time with the squadron.

It has been hard work and we have been at the cutting edge of many things with this new aircraft. The squadron has met the challenges with hard work, innovation, and courage and that is how we are preparing for our first overseas deployment, namely to Japan.

We concluded by reflecting on the history of the Green Knights who from the beginning brought innovation to the fight in the Pacific.

Historically it is interesting to note that VMF-121 was activated in June 1941 and began flying air ground combat missions in August 1942, with the "Cactus Air Force" on Guadalcanal. The Green Knights made Marine aviation history with fourteen aces, including the legendary Joe Foss CMH so the F-35 enabled squadron is making its own aviation history.

Recently, the Vietnam generation Green Knights visited Yuma. Together with the F-35 generation Green Knights, the Vietnam generation Green Knights celebrated the USMC's 241st birthday on November 19, 2016

LtCol Bardo commented: *It was amazing for us to meet with and discuss with the Vietnam-era Green Knights. Although much has changed; much has not.*

What I told the squadron with our visitors present: look at our predecessors and that will be you in a few years. You want to be as proud as they are; to look back at your achievements as being the first F-35 squadron and making aviation history.

You will not focus so much on the hard work we have done over the past two years but will focus on the achievements. And learn from them about how to meet the challenges and serve the nation.

5

The 2018 Visit

From 9 to 10 May 2018, Laird visited MAWTS-1. During that visit, Laird met with the CO of the squadron and talked to him about his perspective on his time at MAWTS-1 and what they accomplished during his time in command.

Col Wellons, MAWTS-1: Shaping a Way Ahead for the USMC and the Joint Force

June 10, 2018

The outgoing CO of MAWTS, Col Jim Wellons, commented about MAWTS-1 and military transformation as follows:

With the coming of the F-35, the Marines have led the way at the outset for the U.S. services which has meant that the Marines have been working closely with the USAF as that service brings its F-35s into initial operating capabilities. We have always had a close relationship with the U.S. Navy. We are after all Naval aviators. I cannot over-emphasize our close working relationship with the U.S. Navy and Top Gun, where we have always had several USMC aviators filling highly sought after exchange tours.

We have some challenges but also many opportunities. Top Gun has a strong emphasis on Super Hornet and are just beginning to roll out their F-35C course, which we intend to support. We have legacy F/A-18s but do not fly the Super Hornet and the USMC has been leaning forward on the establishment of the full spectrum of F-35 tactics, having already executed five WTI classes with the F-35B.

Recently we have made huge strides in establishing ASLA joint communications standards and we are closer now than ever before to aligning all the service standards with joint communications – all the service weapons schools have been cooperating in this effort.

Col Wellons at MCAS Yuma. Credit Photo: USMC

With regard to working with the USAF — over the past decade, as we operated together during the wars in Iraq and Afghanistan, we became much closer and better integrated across the service weapons schools. But the advent of the F-35 has really accelerated our close working relationship with the USAF.

The standup of F-35 was "joint" from the very beginning, and the USMC has been aggressive with the stand up of our operational F-35s – the first of all the services to declare IOC, deploy overseas, and conduct weapons school courses with the F-35. As a result, we have been at the forefront of lessons-learned with the aircraft in terms of sustainment, deployability, expeditionary operations and tactical employment.

We currently have a former USMC F/A-18 instructor pilot flying F-35As on an exchange tour with the USAF Weapons School, and we will soon have the first USAF F-35 exchange pilot coming to Yuma for a tour as instructor pilot in the F-35 division at MAWTS-1. We are all learning about employing, supporting and sustaining the F-35, and deploying it to places like the Western Pacific, where VMF-121 has been in place now a year.

Question: During my time in Australia earlier this year, the Commander of the 11th Air Force raised a key question about the need for the USAF to ramp up its mobile basing capabilities.

How has the USAF interacted with the Marines at Yuma with regard to working through a new approach?

Col Wellons: *Within the USMC, expeditionary operations are our bread and butter. In a contested environment, we will need to operate for hours at a base rather than weeks or months. At WTI we are working hard on mobile basing, and, with the F-35, we are taking particular advantage of every opportunity to do distributed STOVL operations. It is a work in progress but at the heart of our DNA.*

We will fly an Osprey or C-130 to a FOB, bring in the F-35s, refuel them and load them with weapons while the engines are still running, and then depart. In a very short period of time, we are taking off with a full load of fuel and weapons, and the Ospreys and/or C-130s follow close behind. We are constantly working on solutions to speed up the process, like faster fuel-flow rates, and hasty maintenance in such situations. Of course, we have operated off ships with our F-35s from the beginning, and that is certainly an expeditionary basing platform.

We have been pleased with what we have seen so far in regard to F-35 readiness at WTI. For example, in the last WTI class we had six F-35s and we had five out of six up every day, which was certainly as good as anything we have seen with legacy aircraft.

During the most recent class, F-35s supported us with over 95 sorties and a negligible cancellation rate. Our readiness rates at WTI are not representative of the fleet, where we continue to work on enhancing overall readiness goals with F-35.

We then discussed the F-35 and USMC operations beyond MAWTS-1.

Col Wellons: *This is still an early variant of this airplane. It is the early days for the F-35, and we are working things like software evolution. Yet the aircraft has already had an impact in the PACOM AOR. We can put this airplane anywhere in the world, sustain it and fly it and get a key deterrent impact, as we have already begun to see.*

Question: Looking back at your two and half years in command at MAWTS-1, what are some of your thoughts about the dynamics of change which you have seen while here?

Col Wellons: *When I came here, the squadron was in great shape. I had the impression that what I needed to do was to focus on trying to sustain the standard of excellence that had already been established, because the squadron was really firing on all cylinders. I felt we were training at a world-class level and were training to the appropriate skills.*

But during my first year we faced dramatic and significant readiness challenges across Marine aviation, almost at an historic level. This led to significant reductions in the level of pilot proficiency and material readiness and challenged our ability to meet training objectives during WTI.

The readiness cratering also impacted morale and placed our staff in a difficult position. If you have students that are coming to WTI that are barely qualified, who have just barely achieved the prerequisites necessary to come to a WTI class, that creates a risk management problem and makes it difficult to train at the graduate level.

We were looking at dips in proficiency from flying 15-20 hours a month down to 10 or 11 hours a month or lower, and this required us to make some substantial adjustments to how we approached and ran the WTI class. Fortunately, this situation has dramatically changed for the better.

During this last WTI course we had the highest level of readiness that I think we have ever seen for our fixed wing fleet, and our pilots are back above 20 hours a month across all communities. I would caution that we view this readiness recovery as fragile at this point, but it is definitely headed in the right direction.

Question: Clearly, there is a strategic shift underway for U.S. and allied forces to now operate in contested environments. That has happened during your time here.

How has that affected what you have had MAWTS-1 focus upon during your time at MAWTS-1?

Col Wellons: *The team at 29 Palms as well as at Yuma have ramped up the contested and degraded environment that we present to our training audience at WTI and across all the other service level MAGTF training venues.*

We have challenged our students, especially this year, to operate in environments where communications and navigation systems are challenged, facing the most sophisticated and capable adversaries we can find.

We focused on the idea that in the future fight our primary means of navigation and communication will probably be denied, and certainly degraded and our operators may have to use old fashioned techniques to get bombs on target.

Question: *You are clearly working what might be called F-35 2.0 while flushing out the dynamics of 1.0.* And one key area where that is happening is with regard to the sensor-shooter relationship. We talked last year about this dynamic, what has happened since then?

Col Wellons: *In part, it is about the transformation of the amphibious fleet whereby the shipboard strike systems or sensor systems can work with the reach of the F-35 as a fleet.*

For example, we see clear interest from the Navy's side in exploiting 5th generation capabilities in the amphibious fleet using the Up-Gunned Expeditionary Strike Group (ESG), that will better leverage the capability they have got with the F-35.

Naval integration will be a major line of effort in the WTI course going forward.

The F-35 is leading to a fundamental reworking of where we can take the sensor-shooter relationship. We tend to focus on the airplane's sensor and how that sensor can go out and find a target and employ its own ordinance on that target. That is certainly something which the F-35 can do.

But it can also enable an off-board shot, as in the case of HIMARS/F-35 integration. Or it can work with the G/ATOR radar on the ship or the ground to enable weapons solutions for other platforms in the distributed battlespace.

It then becomes a question of how do I maximize the number of targets I can hit with the F-35 distributed force rather than how many targets can an individual fighter hit. This is part of the combat learning we are working on at MAWTS-1 as well.

Question: Assuming readiness remains at an appropriate level, could you discuss the challenges which you see in the near term with regard to training.

Col Wellons: *Clearly, a major challenge we face is the limitations of our training ranges. We need to expand the potential of tasks we can do on these ranges to replicate a realistic and lethal contested environment.*

This is another consequence of our budget challenges in recent years, and we are pushing hard for upgrades of all our emitters, target sets, and simulation capability in order to enable full spectrum training at the high end.

MAWTS-1 Works Change: LtCol Ryan Schiller and His Team Discuss the Way Ahead

July 10, 2018

LtCol Schiller. the Aviation Development, Tactics and Evaluation Department Head and his team of LtCol Waldron, Maj Watts, Maj Zasadny, Maj Buxton, and Capt Jacobellis discussed technology and adaptations they were working with regard to the MAGTF.

With regard to TACDEMOs, among the key efforts were the use of the Light-Marine Air Defense Integrated System (L-MADIS) for force protection/defense, the use of the new Gorgon Stare ISR system, improvements such as using new noise cancelling helmet microphones and new technologies for JTAC cueing, etc.

With regard to TTPs Development, a number of initiatives were conducted which allow the Marines to operate with more lethality and survivability in a variety of combat settings.

These TTPs included working with a new precision penetrator warhead on the laser guided APKWS rocket, F-35 distributed STOVL operations, F-35 and HIMARS integration as well as various aspects of MAGTF digital interoperability such as operations in a contested environment.

MAWTS-1 is working with industry and various USMC and USN organizations to test out new technologies as well as new TTPs to support the strategic shift from a focus on counter-insurgency operations to contested near-peer force-on-force engagements.

According to LtCol Schiller, a key function of ADT&E is to assist in the process of informing future requirements.

It is part of our mission to help requirement officers in Headquarters Marine Corps. We do this by taking items from DARPA, research labs, industry and the PMAs and integrate them into WTI courses. We then provide an after-action report with our assessment on their performance and utility to the force.

The students and instructors in the course help provide a realistic performance assessment for new equipment or technologies in an effort to help the USMC fill current and future warfighter gaps.

A particularly interesting TACDEMO was JTAC Virtual Cueing. In effect, this piece of equipment is part of the shift to enable increased situational awareness to the JTAC. It also has the potential to improve training.

JTAC Virtual Cueing allows for us to use simulated threat environments, close air support aircraft and weapons in an objective area that can measurably increase training and proficiency of JTACs while significantly reducing costs to the taxpayer.

However, solely focusing on the GCE may result in missing some of the less visible aspects of what is going on at MAWTS-1. Through a variety of developments, the Marines are focusing on extending the range of offensive and defensive capabilities in the battlespace from the air and sea to support the GCE ashore.

While the Marine Air Ground Task Force tends to operate as its own entity, these new technologies will permit synergistic lines of effort with joint or even allied ground, air and naval elements.

One example of this being tested at MAWTS-1 is the continued integration of Ground/Air Task Oriented Radar (G/ATOR) TACDEMOs. G/ATOR provides targeting information and fires support ashore.

One challenge will be to shape a MAGTF, joint and allied understanding of how to efficiently operate in concert. This is magnified with the introduction of the F-35 which provides significant MAGTF organic support but also possesses capabilities to enable joint and allied fourth generation aircraft as well.

A key focus of efforts during recent WTIs was how to manage and conduct Integrated Fire Control between G/ATOR and F-35.

As was commented in the discussion: *We focused on how to maximize three core systems – G/ATOR, the Composite Tracking Network (CTN)* and *CAC2S (the latest software iteration of the Common Aviation Command and Control System) as they are fielded to the force for the first time as a systemic whole.*

We are going to be able to provide significantly greater information to all of the shooters, whether airborne, shipborne or ground based missile defense systems.

A key challenge for the ADT&E Department at MAWTS-1 is to change the mindset of Marines in order to get them to understand and adopt new TTPs for new systems and not simply adopt a less effective legacy mindset.

In addition to TACDEMOs, the ADT&E Department continuously refines TTPs for the Fleet. One TTP initiative conducted during the WTI course was Distributed STOVL operations (DSO) with the F-35, which is clearly a work in progress.

During the previous WTI they conducted DSO with MV-22 support, but during this course they did used a KC-130J.

It was noted in the conversation: *The KC-130J took off with Marines, ordnance, fuel and a security team. They landed at a remote facility, set up two forward arming and refueling points as well as a defensive perimeter.*

Shortly thereafter, two F-35s landed and received hot fuel and hot-loaded ordnance, then they took off and executed their mission. We are clearly working towards and expeditionary air base type of mindset for the force.

They also integrated TTPs for both Group 4 and Group 1 UASs, both in an offensive capability as well as in a layered defense from threat UASs.

Finally, we discussed digital interoperability and its role in the evolution of the MAGTF. And Laird had a chance to see and work with the MAGTAB. This is a commercial tablet with an encrypted link to provide a means for the MAGTF to handle the transfer of relevant data throughout the Aviation Combat Element and Ground Combat Element (GCE).

The Marines have taken an off-the-shelf commercial technology and adapted it to provide core data communications capability within the USMC, and as one Marine put it: *We have shown others in the joint force that you don't have to write a complicated requirements document to get a cutting-edge capability.*

As was noted: *This represents significant progress in terms of understanding how we can leverage commercial technology in the current fight while still meeting requirements to have low risk in terms of data protection and transmission security.*

During this WTI course, the ADT&E Department placed significant focus on creating a disrupted battlespace which included jamming, electronic warfare, and other key non-kinetic elements.

U.S. Marines with 2nd Battalion, 6th Marine Regiment and Marine Aviation Weapons and Tactics Squadron One (MAWTS-1) conduct digital interoperability training at Marine Corps Air Station Yuma, Ariz., March 28, 2017. Photo by Lance Cpl. Michaela Gregory, MAWTS-1.

Several members of the joint community also participated at WTI bringing advanced capabilities to play in the electronic warfare area. The goal is to get Marines to proactively think about how the adversary will conduct battle so that strengths can be countered, and weaknesses exploited.

It is also becoming more and more important to prepare Marines to incorporate the use of non-lethal disruptive technologies and techniques.

One of the MAWTS officers highlighted the fact that when Harriers drop bombs, the pilots and GCE can see the immediate effect. But when they fly jamming pods, they do not see the immediate effect and can be frustrated. As was noted: *But their jamming effect could be much more significant than the bombs that they might have dropped in conflict.*

Learning how to engage in such a manner is part of the technology-training challenge as well. The AN/TPS-80 will replace the AN/TPS-63 and reduces set up time from eight hours to 30 minutes for the system. Marine Air Control Squadron 2 received the first G/ATOR issued to the Fleet Marine Force following testing to improve the squadron's readiness and expeditionary capabilities.

MAWTS-1 Works F-35 Integration: The Case of HIMARS

July 10, 2018

MAWTS-1 is working on reshaping Marine Corps approaches to moving forward from a primary focus on counter insurgency. The shift is a significant one, which requires reshaping approaches, leveraging new capabilities, and integrating those capabilities into the overall evolution of the MAGTF. It is a work in progress, and one driven by technology, combat experience and cross-learning from other U.S. services as well as allies.

An example of the work in progress is providing a capability for an advanced ground-based strike missile to operate with greater lethality when guided by a low observable air combat system which identifies targets beyond range of sight and not identified by the systems organic to that strike system. This is an example of how the sensor-shooter relationship needs to evolve when operating at greater distance and in a contested battlespace.

The core approach is to find ways to leverage the F-35 to provide an expanded aperture of support for the Ground Combat Element when executing the ground scheme of maneuver in a peer-to-peer conflict. As the F-35 operates in its low observable mode and generates through its DAS and integrated sensors a battlefield situational awareness 'map,' targets can be identified deep within the enemy's operating area. Targeting information can be generated to the Marines or to other joint forces to provide for precise fires targeting. It is clear that the F-35 has an extraordinary sensor capability and sensor system integration, which can empower C2 in the operational battlespace.

In visits and discussions we have had with allied air forces flying the F-35, the use of the new systems was already evident. In one case, an Air Force was using sensor data from its aircraft to provide significant SA to that ally's navy as well as other capabilities for the fleet as well.

In another case, an ally is flying a single F-35 along a border where low flying threats are crossing the border regularly with drugs, weapons, and other undesirable deliveries onto that ally's territory. The F-35 is providing coverage of the entire border area and delivering that information including guiding border forces to mission success. The aircraft's ultimate benefit is to provide a major contribution to information dominance.

The Marines are clearly among the most inventive of forces in pursuing ways to leverage the F-35 as a multi-domain flying combat system. But this is not simply going to happen without work of the aviation with the ground communities working closely together as they do at MAWTS-1.

For the Marines, working F-35 integration with HIMARS was worked during WTI-2-18 and Laird had a chance to discuss the way ahead with Major "Doctor" Buxton, MAWTS-1 Air Office Department, Major Andrew Crist, Fixed Wing Offensive Support Specialists, and Major Joshua Freeland, a Direct Air Support Control Officer.

What these officers described was a clear work in progress, one which will rely on leveraging software upgrades on the F-35 but concurrent progress with regard to the software and hardware evolutions of the data link systems as well. From this point of view, the F-35, much like the Osprey before it, is playing a forcing function within the USMC for change.

With the Osprey, significant change was driven in how the Marines operated in the land wars, and in how they approached counter-insurgency operations. The F-35 has come precisely at the point when the strategic shift is underway, and it is clear that the U.S. and the allies are using the F-35 as a trigger point for broader transformation as well.

And through this effort, the Marines are looking at broader issues of the F-35 and its role within the overall effort to shape greater digital interoperability for the force as well.

Marines with 5th Battalion, 11th Marine Regiment, 1st Marine Division, arrive at one of their launch positions with the High-Mobility Artillery Rocket System at the Air Combat Element landing strip as a part of Integrated Training Exercise 3-18 aboard the Marine Corps Air Ground Combat Center, Twentynine Palms, Calif., May 21, 2018. Credit: USMC

The GCE fires elements use a data link communication system, which operates by sending what is called K messages. The immediate challenge was to find ways to work the F-35 systems with an ability to work with the data links used by the GCE.

The data links for the GCE are being reworked to be more effective in its operational integration with the Air Combat Element. As the GCE receives new software and hardware systems and as the F-35 evolves to its 3-F configuration an ability to link systems more effectively in the distributed battlespace will be possible.

But the Marines are working the opportunity to do so prior to arrival of the optimal situation. As one Marine put it: *We are looking to build in surface fires capability into the F-35. We started by looking at ways we could use CAC2S as a gateway to enable us to move in this direction.*

CAC2S is the USMC's C2 system designed to provide for integration between the ACE and the GCE. It like the F-35, is a work in progress As the Marine Corps has defined CAC2S:

CAC2S will provide a complete and coordinated modernization of Marine Air Command and Control System (MACCS) equipment. CAC2S will eliminate current dissimilar systems and provide the MAGTF Combat Element with the hardware, software, and facilities to effectively command, control and coordinate air operations integrated with naval, joint and/or combined C2 units.

CAC2S will comprise standardized modular and scalable tactical facilities, hardware and software that will significantly increase battlefield mobility and reduce the physical size and logistical footprint of the MACCS. CAC2S Phase 1 successfully completed its Initial Operational Test and Evaluation (IOT&E) in 2011.

Subsequently, Phase 1 received its full deployment decision on 25 Oct 2011 and limited deployment capability in February 2012. Phase 2 successfully achieved Milestone C decision on 31 Mar 2015 and IOT&E in Apr 2016. A Fielding Decision Review (FDR) was conducted on 11 Aug 2016 and ADM signed on 25 August 2016.[1]

As the Marine Corps gets its updated versions of CAC2S, they are looking to the new capabilities to provide an effective gateway between the message set capabilities of the platforms.

The Direct Air Support Center (DASC) is where the translation and validation occurs on the battlefield and where Link 16 messages from the F-35 would then be translated into K messages for the GCE.

As another Marine put it: *Link 16 J series messages received by the DASC will then be translated into the K series format which the GCE utilizes to generate fire missions and is passed along to fires approval authorities with airspace clearance.*

During WTI-2-18, the Marines used a new VIASAT radio as part of the firing sequence for the F-35/HIMARS tandem. And with a handheld radio able to handle Link 16 messages, and the team was able to use a Link 16 data link from the F-35 to enable a HIMARS firing.

Plaque in MAWTS-1 building commemorating HIMRS-F-35 Integration Effort in 2018.

But it was clear that working the integration was a hard task, one that needs to become much simpler to be effectively operational on the battlefield. Seeing the Marines work the process and in a way that can inform both the upgrade processes on the F-35 as well as with HIMARS and related equipment is impressive.

Clearly, working the data links and communications is a key part of being able to operate on the distributed battlefield. Although a work on progress, it is clearly moving in the right direction.

Integrating the ACE with the GCE in the Re-emergence of Great Power Competition

July 10, 2018

At MAWTS, there was clearly a recognition of the strategic shift facing the liberal democracies and the need to prepare to fight in contested areas with the means which peer competitors can bring to the fight. This clearly is a work in progress, but change is being driven as well by the innovations being introduced by the Marines working the innovative parts of the joint and allied forces as well.

For the Marines, the Osprey has driven significant change in terms of the reach of the Ground Combat Element; now the F-35 is empowering the air-ground team in new ways.

And next up will be the CH-53K, which will both benefit from the transformation of the Marine Corps but also contribute to it as well. And certainly, as MAWTS gets its hands on the CH-53K it will be shaped as a combat platform by the overall transformation of the Marine Corps.

A key element of the transformation is working a digital interoperable force, which can operate, effectively in a contested environment and to do so by highlighting force mobility and capability to operate as a distributed force. All these elements of change were on display at MAWTS-1 during this visit.

In discussions with two leaders of the Ground Combat team at MAWTS, Major Brian Green and Captain Thomas Fields, the role of the Ground-training element within MAWTS was discussed. It is a small part of the overall training of the Ground Combat Element in the USMC and represents a special part of the overall effort. The focus is upon exposing the GCE to the wide variety of air assets which can be brought into to support the GCE both within the MAGTF as well as from the Joint Force.

As Captain Fields put it: *We get a MEBs worth of aircraft for a WTI course, which is something our ground combat forces will not normally see. And they will get exposure as well to the other services air assets, which can be brought to bear on the battlespace.*

They get to understand that capability before they actually might employ it in the future.

The involvement of the GCE has gone through an evolutionary process. According to the officers, it started with the involvement in 1988 of company working with MAWTS-1 a decade after MAWTS-1 had been established.

U.S. Marines assigned to 1st Battalion 4th Marines post security during assault support tactics three during Weapons and Tactics Instructor Course 2-18 in Yuma, Ariz. on April. 20, 2018. U.S. Marine Corps photo by Lance Cpl. Joel Soriano.

They participated but were not well integrated. Then the GCE was upsized at WTI to a battalion level and the Marines began to operate the TALON Exercise, which is a ground combat exercise, concurrently with WTI and thereby providing exercise space for more interactive learning.

With the coming of the new air systems to the Marines, and the clear focus on digital interoperability, this interactive space becomes even more important in shaping concepts of operations and real-world operational capabilities to deliver a more lethal force into the distributed battlespace.

Another part of the challenge is the return to the sea. Major Green has significant at-sea experience, and with the emphasis on the return to the sea this means that the Navy needs to focus on its amphibious warfare skills as the Marines bring new approaches and capabilities to the sea base as well.

And this is clearly a work in progress, because with the decade of the land wars, skills in the amphibious domain have been downplayed and some key skills atrophied. This means the work at MAWTS-1 also needs to take into a account the return to the sea and applying MAGTF integration to the sea base as well.

Another key element of the undertaking is the introduction of digital interoperability as key line of development for the USMC. Clearly, this is not just about the equipment but the soldiers working the equipment and learning how to maximize benefits from digital interoperability as well as to work with the downsides of data as well.

Major Green noted that there is a generational challenge associated with this. *The younger generation are digital natives; the older officers are not. The senior level is wrapping their heads around the transition and working the challenge, but it is a challenge as well.*

An aspect of the training at MAWTS with the ground units, which come into the WTI course, is that they will study the same curriculum as the aviation community for the first two weeks of the 7-week course.

As Major Green put it: *We provide the same read aheads to the infantry as we do for the aviators. And we hold them to the same standards during the first two weeks of the course as well.*

The focus is not upon specialty or MOS training for the ground component, but certification as fire support team leaders. And again, because of the wide panoply of aviation assets the ground force will see during the course and into the exercise, the notion of fire support becomes broadened as new capabilities are seen and understood.

As Major Green noted with regard to GCE involvement in MAWTS-1: *We run two courses a year in conjunction with WTI. We typically have between 10 to 14 students tops.*

The target goal of the effort was highlighted by Captain Fields: *When we receive our ground combat Marines, we will return them to their units in seven weeks and we're going to return a subject matter expert on integration between air and ground assets and how to take a ground scheme maneuver and know how to integrate aviation assets to best support the ground scheme of maneuver in terms of fires, assault support, and sustainment.*

In short, MAWTS-1 provides a venue for the cross training

which makes MAGTF innovation possible. And with the strategic shift away from the counter-insurgency effort, there will be significant demands on the innovation curve.

How to Prevail in a Disrupted and Degraded Combat Environment?

July 6, 2018

In a discussion led by Captain Michael Jacobellis, the Ground-Based Air Defense Division head at MAWTS, the challenge posed by adversary jamming and other EW challenges which disrupted the Marine's normal C2 and ISR data links and communication tools was discussed.

Clearly, part of that challenge is learning what is simply equipment malfunction versus a deliberate adversary disruptive strategy. Part of it is learning to work with a diversity of targeting and communication tools to do work arounds when disruption comes.

Some of the new equipment is clearly designed to provide a combat advantage for the Marines, such as the new G/ATOR radar which provides targeting tracking information which can inform the air element as well as work with the ACE to provide redundant capabilities.

As one Marine noted with regard to G/ATOR: *We are working with a family of systems that will allow us to provide a recognized air picture across the entire MAGF so we can pass targeting data across the MAGTF as well.*

This system is the Composite Tracking Network or CTN. This was the first time we failed G/ATORS at WTI and they allow us through use of CTN to integrate all of the sensors together and empowers integrated fire control.

The shift from counter-insurgency habits, equipment and operations is a significant one and is clearly a work in progress. It is about shedding some past learned behavior and shaping more appropriate ways to operate as a force in a contested electronic warfare environment.

And disruptive technologies, which the adversary might use

against the Marines, were being fielded to test the USMC approach. One example is the use of multiple drones or UAVs against Marine Corps forces and testing various technologies and approaches to attenuating that threat.

The Marines are working closely with the U.S. Army on this effort, and our visit earlier this year to Fort Sill highlighted progress, which the U.S. Army is making with regard to fielding a new capability to deal with UAVs disrupting the battlespace.[2]

Similar to the Army, the USMC is working new systems onto a combat vehicle to shape more effective ways ahead as well. The USMC system is called L-MADIS or the Light Marine Air Defense Integrated System (L-MADIS) which is designed specifically for counter UAS missions.

It is a two-vehicle system which works the ISR data, and C2 links and delivers a counter strike capability against incoming UAS systems. The L-MADIS system is very expeditionary and can be carried by MV-22s or C-130s. The Army's version is being built off of a Stryker vehicle, and the Marines off of a JLTV vehicle.

The same instinct is in play – use a core vehicle in use for the ground forces, shape a flexible management system on the vehicle and have modular upgradeable systems providing what BG McIntire at Fort Sill referred to as the "toys on top of the vehicle."

In other words, combat learning can shape the systems being put onboard the vehicles and working commonality with the US Army can provide for a broader deployed capability dependent on how the force will operate or build up in an objective area.

The Marines are building the ground vehicle and systems infrastructure within which they can plug evolving counter-battery fires capabilities as those develop. And clearly, they are looking at extended range as well for the counter-battery fire.

Clearly, one aspect of the combat learning has been that Army and Marine Corps ground forces need to tap into similar capabilities when they have them to provide for enhanced joint capabilities.

The C-RAM opportunity is clearly one of them. And generally, active defense capabilities have been highlighted again within combat preparations.[3] The Marines rely on an upgraded Stinger missile and are looking forward to the introduction of directed energy weapons, again working closely with the US Army.

VMX-1 Working a Way Ahead for the MAGTF

July 10, 2018

With the coming of the F-35, VMX-22 has become VMX-1, and has generated a broadening of the aperture of what the new combat air assets can do in the process of the transformation of the MAGTF and its reach back as well. The CH-53K will soon come into this transformation process and will both contribute to and leverage the overall process of change.

During the visit to Yuma in May 2018, several members of VMX-1 during a roundtable discussion provided an update on the Osprey, on the ongoing work with regard to the F-35B, and the core effort to shape a digitally, interoperable MAGTF.

Currently, VMX-1 owns six F-35Bs with two in depot maintenance at Cherry Point and second undergoing modifications at Edwards AFB. Two of the officers at the roundtable have extensive experience working with the aircraft.

Major Brendan Walsh who has been flying F-35Bs since their standup at Eglin AFB provided an overview on the aircraft within MAGTF operations. He served as the operations officer for the Green Knights and worked their preparation for deployment to Japan.

He is currently leading the Marine Operational Test & Evaluation Squadron 1 F-35B Detachment at Edwards Air Force Base, CA., under the command of Col Rowell. Major Walsh had just under 600 flight hours in the F-35B when we discussed his work during the roundtable.

A second Marine who discussed the F-35B from a strong basis of experience working with the aircraft was Major Paul Wright. Major Wright was originally a F-18 pilot but has worked through training and related F-35 activities to become a test pilot with VMX-1.

Major Walsh worked DT-3 as well where 12 F-35Bs operated with Ospreys onboard the USS America and supported assault operations in the San Clemente Island test area.

They underscored that the Osprey working with the F-35B enabled an ability to insert force into hostile areas of the sort being prioritized now with the strategic shift underway.

Given Laird's recent visit to the UK and discussions at Marham and Portsmouth, he discussed with them their working relationship with the Brits. Not surprisingly, the two pilots emphasized the close working relationship but in so doing underscored a core point about the F-35 operating community worldwide.

As Maj. Wright put it: *I see the Brits on a daily basis at Edwards.*

It should be noted that at Edwards all three U.S. services flying the F-35 as well as the Brits and Dutch work with the USAF Test Squadron.

And as Maj Walsh added with regard to their work experience at Edwards: *We fly together regularly and work together closely as well. It is great to work commonality and to understand differences as well with how the partners and the services are working the airplane as well.*

The aircraft is a flying combat system with significant C2 and ISR capabilities. And a clear challenge is task management with the aircraft as opposed to sensor management, which is what a legacy multi-mission fighter will focus upon. As a multi-domain fighter the focus of the information generated and displayed in the cockpit is about allowing the pilot to task manage and enable other fighters to be more lethal and survivable.

As Major Walsh put it: *With legacy platforms you would have one piece of the situational awareness puzzle and have to rely on other platforms to direct you. And we elevate the team operating around us.*

Major Wright added: *There is a culture or mindset change for the F-35 pilot compared to a legacy pilot. You work with the key information for the combat situation and can choose to employ your weapons, both lethal and non-lethal, against a target, or pass that information on and employ another asset in the attack phase of the OODA loop.*

Major Walsh then added with regard to his Red Flag experience in 2016: *If you look at what Vipers can achieve with and without the F-35, the difference is dramatic. You will find a lot of legacy platforms very happy to operate with F-35s for sure.*

At MAWTS-1, they are working the STOVL distributed operations piece as well. They have worked with the MV-22 providing the support last year, and at the last WTI, they used the C-130. VMX-1 is clearly involved with shaping the TTPs for this operational capability.

The Marines initially worked the IOC support for the F-35, now they are working on leveraging the aircraft in its current configuration for the MAGTF and will spend the next couple of years working with the next iteration of the aircraft, the 3F software configuration and winging out its capabilities, notably for the Ground Combat Element.

The 3F will allow the aircraft to work with externally loaded ordinance for operational situations in which Low Observability is not the primary aspect of what is being required from the aircraft. And while they are doing that, in the words of Major Wright, there will be a lot of "side projects' regarding how to best leverage the F-35 for the MAGTF as well.

With regard to the Osprey, the air system has evolved from the VMX-22 role of getting the aircraft into the force, to reworking the con-ops of the MAGTF leveraging the aircraft, to modifying aspects of the aircraft and its operations to optimize force insertion, to reshaping the aircraft to become a key part of the digital interoperability effort within the USMC.

Major Duchannes and Maj Ryan Beni, also an MV-22 operator since 2009, and a Marine Laird had met earlier during a visit to Marine Corps Air Station New River, noted that the MV-22 has gone through several upgrades over the past few years, including upgraded communication systems and a new defensive weapons system.

We are now working our way into the whole digital integration realm, so that we can empower the MAGTF more effectively as well as the Joint Force.

The Osprey pioneered what might be called digitally enabled force insertion with the introduction of the MAGTAB into the squads operating on the Osprey and having the situational awareness to understand what had changed in the objective area during a long flight on Osprey to the objective area as well.

And working that digital interoperability piece is crucial for the F-35 as well as the Marines work the gateways and message systems to get benefit from the F-35 as it operates as a forward combat system enabling MAGTF operations.

Raising the Bar on Maintainability

July 10, 2018

Maintainability is a key aspect to ensure aircraft availability to support combat operations. The readiness crisis of the past few years has also included the challenge of upgrading maintenance support efforts as well.

The strategic shift from counterinsurgency to preparing for force-on-force conflict and a more rapid battle rhythm affects the maintenance side of the force as well. The wars in the desert have been tough on the aircraft, and pushing out their operational life has been a challenge. But with a more rapid pace of combat as a key focus of attention, how to ensure better aircraft readiness, and fleet availability?

It is difficult to get a fleet approach if there is not commonality among the aircraft and understanding of best practice standards. This will not happen without shaping the most effective standard practices for maintainers in supporting aircraft and ensuring that there is an effective rhythm between maintenance to support daily operations and ensuring that a maintenance cycle that provides for aircraft to have regular maintenance that ensures longer performance cycles as well.

One way to address this is to ensure that best practices can be established throughout the maintainer force and have an evolving understanding of the best standards to achieve maintenance of each type model and series of aircraft.

For the Marines, a core focus is upon training maintainers who can work across a range of aircraft. The Marines are focused on expeditionary operations, and mobile basing.

And this means, the Marines need maintainers who can support aircraft that operate from the mobile base, and not aircraft that need to go to a "Walmart" type supply base to be maintained by maintainers clearly divided into stove piped specialties.

And the introduction of new aircraft, ones that are software upgradeable, is clearly changing the system as well. The coming of the F-35 is driving change which needs to be managed into shaping effective standardization as well. The F-35 highlights another change: the growing importance of the software management side of the equation.

In recognition of the central role, which enhanced maintainability, plays for the USMC in the context of the strategic shift a new course was brought to MAWTS last year. This new course and its role was well described in an article by Captain Harley Robinson published on March 15, 2017 by the 3rd Marine Air Wing.

A new course has been added to this year's Weapons and Tactics Instructor Course (WTI), held at Marine Corps Air Station Yuma, Arizona.

Advanced Aircraft Maintenance Officers Course (AAMOC) is a second-level graduate school for professional aircraft maintenance officers in the Marine Corps.

After the initial school for aircraft maintainers at Naval Air Station Whiting Field, Florida, there are limited follow-on training opportunities. AAMOC is the first of its kind and the expectation is to increase standardization, improve aircraft readiness and to minimize aircraft mishaps.

Our primary school we go to is a Navy school, and that program is extremely successful for the Navy and its officers, but when Marines graduate, we train slightly different once we leave the school," said 1st Lt Jared Hasson, an assistant aircraft maintenance officer with Marine Aviation Weapons and Tactics Squadron (MAWTS) One and AAMOC instructor from Winter Haven, Florida.

In an ever-evolving job field, the course's main purpose is to create a higher understanding and standardized learning platform for professional maintenance officers.

MAWTS-1

There are training gaps with what is taught in the schoolhouse and what is actually being done in the fleet," said Capt. Scott Campbell, an a chief instructor and AAMOC developer from Amarillo, Texas. "This class is an attempt to formalize, consolidate and structure information that goes into the fleet that isn't getting taught to the Marines."

The curriculum consists of an initial and final exam, roughly 62 hours of academic course work and additional training outside the classroom. There will be daily evaluations of the students by the instructors on class work, practical application, and projects. The students will receive grades on every subject and must maintain an 80% grade average to graduate. The course will run a total of seven weeks.

If every person in the classroom is able to walk away with something they didn't know beforehand, then I would deem this a success," said Campbell. "This isn't going to immediately stem the flow or in no way is designed to be the sole thing that fixes aircraft readiness. But teaching our maintenance officers how to better utilize their aircraft, where the demand of the aircraft comes from and how to manage that, absolutely contributes to better readiness numbers."

Graduation is scheduled for April 30, which is the end of WTI. Graduates from AAMOC will be granted signing authority for 2000 level codes in the newly minted T&R manual and will have the title of Maintenance and Training Instructors (MTI).[4]

During my visit to MAWTS-1, I had a chance to meet with Captain Campbell and Lt Hasson and to discuss this fleet focus. It was quite an experience.

I have never encountered more enthusiastic maintainers in my life, and listening to them it was clear that being a maintainer for today's Marine Corps is a worthy calling in service of the nation. It would be difficult for me to convey the sense of enthusiasm, passion, and commitment these two Marine leaders provided in our discussion.

Suffice it to say that the sense of urgency in getting the readiness upsurge and the re-set of maintenance standards is a core part of reshaping the force and getting ready for the next fight.

Their core effort is upon reshaping the culture of maintenance in the USMC. They are focusing on what they can do at their level and have targeted their efforts on enhanced training and effectiveness of those Marine Corps officers who are training the maintainers. By so doing, they are generating a ripple effect throughout the maintenance culture of the USMC itself.

According to Captain Campbell: *What is our objective with The Advanced Aircraft Maintenance Officers Course? What we're doing is standardizing the MOS. Because I can't control, at my level as a company-grade officer, contracts.*

I'm not going to be able to control procurement. So what can I control?

We are focused on how we do things; how we share knowledge about best practices and how we keep from having to keep reinventing the wheel.

He highlighted the inventive quality of Marines, which is a plus, but the downside is that unshared individual innovations will not drive overall change unless knowledge is shared and best practices determined.

And the course is taught in such a manner that whatever the baseline of the course going in, it is altered in the interactions with the students and articulation of best practices.

For the WTIs a large number of Marine Corps aircraft come to the course, to the exercise. This provides an opportunity to bring significant experience from throughout the Marine Corps with different aircraft concentrated in Yuma.

And this provides a significant learning opportunity. As Captain Campbell put it: *The maintainers come to Yuma as if on a deployment. They are in barracks together and they generally don't know one another. They cross learn during their time at WTI and with regard to the course, those participating in the course will shape a cross learning network which informs the course and provides an interactive baseline as we push forward effective standardization.*

1st Lt. Jared Hasson underscored that in the class maintainers who are new to the game as well as those who have written the book with regard to particular aircraft.

At the outset, the gray beards look around and ask why are the newbies here? We wrote the book. But then the newbies challenge the way things are done and look for new ways to do things and soon a cross learning process is underway.

The course is part of an effort to generate broader understanding throughout the maintenance community of best practices and how to work from the evolved best practices to generating further progress.

Screenshot taken from USMC video, 2018.

An example cited by Captain Campbell was with regard to the F-35. He worked earlier on the F-35 and one of the concerns was how to get the F-35 maintenance system to plug into and work with the broader Marine Corps Maintenance IT system.

He noted that the Green Knights are currently deployed at sea, and they are sorting this out. His task then is to find out how they have done it and move that learning into the course and then share the knowledge so that the Marines do not have to have one-off learning efforts. They want to shape a learning curve.

Captain Campbell also underscored that the coming of the F-35 highlighted another change: *More of our maintenance is going to come from software and avionics than anything else.*

1st Lt. Jared Hasson highlighted another aspect of their work at MAWTS-1, namely generating Capstone projects. These projects are generated by the students to develop point papers on a subject of their interest and work through ways to deal with a problem of their own choosing.

They figure out how to address the problem; and they work on that throughout the course. And at the end of the course, they present their capstone project findings and some of these projects come to the attention then of the Naval Postgraduate School which then further pursue them.

And shaping best practices is a key element for enhancing readiness and maintainability of the ACE. Bringing a new focus within MAWTS-1 to this challenge certainly will help the overall efforts to transform the force going forward.

6

2020 Interviews and Visit

The significant change in USMC aviation since the introduction of the Osprey has set in motion fundamental changes overall in USMC capabilities and concepts of operations. In the past decade, the Osprey has matured as a combat platform and fostered significant change in concepts of operations.

No less than the virtual end of the ARG-MEU and the shaping of a new approach to amphibious warfare and shaping new concepts of operations for dealing with peer competitors is underway.

With the end of the primary focus upon the land wars, the Osprey and changes to the attack and support helicopter fleets, have changed how the Marines can operate in a combat space. The revolution in tiltrotor technology, and the much more effective integration of the Yankee and Zulu class helicopters, have allowed the Marines to have a smaller logistical footprint in covering a wider combat space.

This revolution in tiltrotor technology is broadening to the joint force with the Navy adding the CMV-22B and the Army having decided to replace the Blackhawk with the FLRAA or the Future Long Range Assault Aircraft.

Enter the F-35B. With the coming of the F-35B and the impact of the template of change laid down by the Osprey, with its range and speed, together they are driving significant change in distributed operational combat capability. This capability has been not only reinforced but is being taken to the next level.

Now with CNI-enabled aircraft with 360-degree situational awareness, a Marine Corps MAGTF can deploy with an integrated EW-ISR-C2-weapons carrier and trigger which can form the backbone for enabling an insertion force.

What Laird saw in his 2020 interviews and in his visit to MAWTS-1 that year was how a command focused on real world force integration and warfighting began the process of change from a primary focus on the land wars to the return of great power competition. And the use of I refers to what Laird wrote about the 2020 visit in the various interviews.

USMC Aviation Innovations for the 2020s

The projected additions of USMC aviation assets in the decade ahead clearly can provide key capabilities to enable this transition, much like the changes of the past decade put the Marines into this position in the first place.

Three key additions are crucial to this evolution.

1. The Addition of the CH-53K

The first is the addition of the CH-53K. Without an effective heavy lift asset, an ability to operate from the seabase or to establish distributed FARPS in the operational window for an integrated distributed force, it will be difficult to have a sustainable distributed force.

The CH-53K provides a key element of being able to carry equipment and/or personnel to the objective area. And with its ability to carry three times the external load of the CH-53E and ability to deliver the external load to different operating bases, the aircraft will contribute significantly to distributed operations.

The digital nature of the aircraft, and the configuration of the cockpit is a key part of its ability to contribute to support of the force as well. The aircraft is a fly-by-wire system with digital interoperability built in. With multiple screens in the cockpit able to manage data in a variety of ways, the aircraft can operate as a lead element, a supporting element or a distributed integrated support node to the insertion force.

A key change associated with the new digital aircraft, whether they are P-8s or Cyclone ASW helicopters, is a different kind of workflow. The screens in the aircraft can be configured to the task and data moved throughout the aircraft to facilitate a mission task-oriented workflow.

In the case of the CH-53K, the aircraft could operate as a Local Area Network for an insertion task force, or simply as a node pushing data back into the back where the Marines are operating MAGTBs.

Marines carrying MAGTBs onboard the CH-53K will be able to engage with the task force to understand their role at the point of insertion. The CH-53K as a digital aircraft combined with the digital transformation of the Marines create a very different ground force insertion capability.[1]

2. The Addition of Capable Unmanned Assets

The second is the addition of new and more capable unmanned assets to empower the force, and to provide for the proactive ISR which the integrated distributed force needs to enhance their operational effectiveness.

The remotely piloted unmanned assets are being added to the force in the first part of this decade, such as the Reaper, the second part of the decade will see the emergence of operational autonomous systems, below, on the sea and in the air enabling a more lethal and survivable distributed force.

3. Progress Shaping the Digital Integration of the Force

The third is further progress in shaping the digital integration of the force so that distributed operations can be more effective in contested environments. The significant changes in C2 and ISR capabilities, integration and distribution is many ways the 6th generation rather than being a new aircraft.

For the Marines, working digital interoperability has been a high priority as they prepared for the shift from the land wars to engaging in contested multi-domain operations. At MAWTS-1, innovation is driven by the work done in the WTIs, not in wargaming. It is to the role of MAWTS-1 in 2020 to which we now turn.

Colonel Gillette, CO of MAWTS-1, 2020

September 22, 2020

During Laird's visit to MAWTS-1 in September 2020, he had a chance to engage with a number of MAWTS-1 officers and with the CO of MAWTS-1, Colonel Gillette, with regard to the USMC's emphasis on their contribution to naval warfare.

Question: How is the Marine Corps going to contribute most effectively to the Pacific mission in terms of Sea Control and Sea Denial?

And how to best contribute to the defensive and offensive operations affecting the SLOCs?

Both questions highlight the challenge of shaping a force with enough flexibility to have pieces on the chess board and to move them effectively to shape combat success.

Col Gillette: *Working through how the USMC can contribute effectively to sea control and sea denial for the joint force is a key challenge.*

The way I see it, is the question of how to insert force in the Pacific where a key combat capability is to bring assets to bear on the Pacific chessboard. The long-precision weapons of adversaries are working to expand their reach and shape an opportunity to work multiple ways inside and outside those strike zones to shape the battlespace.

What do we need to do in order to bring our assets inside the red rings, our adversaries are seeking to place on the Pacific chessboard?

How do you bring your chess pieces onto the board in a way that ensures or minimizes both the risk to the force and enhances the probability of a positive outcome for the mission?

How do you move assets on the chessboard inside those red rings which allows us to bring capabilities to bear on whatever end state we are trying to achieve?

For the USMC, as the Commandant has highlighted, it is a question of how we can most effectively contribute to the air-maritime fight.

For us, a core competence is mobile basing which clearly will play a key part in our contribution, whether projected from afloat or ashore. What assets need to be on the chess board at the start of any type of escalation? What assets need to be brought to bear and how do you bring them there?

I think mobile basing is part of the discussion of how you bring those forces to bear.

How do you bring forces afloat inside the red rings in a responsible way so that you can bring those pieces to the chess board or have them contribute to the overall crisis management objectives?

How do we escalate and de-escalate force to support our political objectives?

How do we, either from afloat or ashore, enable the joint Force to bring relevant assets to bear on the crisis and then once we establish that force presence, how do we manage it most effectively?

How do we train to be able to do that?

What integration in the training environment is required to be able to achieve such an outcome in an operational setting in a very timely manner?

Question: Ever since the revival of the Bold Alligator exercises, I have focused on how the amphibious fleet can shift from its greyhound bus role to shaping a task force capable of operating in terms of sea denial and sea control.

With the new America-class ships in the fleet, this clearly is the case. How do you view the revamping of the amphibious fleet in terms of providing new for the USMC and the U.S. Navy to deliver sea control and sea denial?

Col Gillette: *The traditional approach for the amphibious force is move force to an area of interest. Now we need to look at the entire maritime combat space, and ask how we can contribute to that combat space, and not simply move force from A to B.*

I think the first leap is to think of the amphibious task force, as you call it, to become worked as key pieces on the combat chess board. As with any piece, they have strengths and weaknesses.

Some of the weaknesses are clear, such as the need for a common operational picture, a command-and-control suite to where the assets that provide data feeds to a carrier strike group are also incorporated onto L-Class shipping.

We're working on those things right now, in order to bring the situational awareness of those types of ships up to speed with the rest of the Naval fleet.

Question: A key opportunity facing the force is to reimagine how to use the assets the force has now but working them in new innovative integrable ways or, in other words, rethinking how to use assets that we already have but differently.

How do you view this opportunity?

Col Gillette: *We clearly need to focus on the critical gaps which are evident from working a more integrated force. I think that the first step is to reimagine what pieces can be moved around the board for functions that typically in the past haven't been used in the new way.*

That's number one. Number two, once you say, "Okay, well I have all these LHA/LHD class shipping and all the LPDs et cetera that go along with the traditional MEU-R, is there a ship that I need to either tether to that MEU-R to give it a critical capability that's autonomous? Or do I just need to have a way to send data so that they have the same sensing of the environment that they're operating in, using sensors already in the carrier strike group, national assets, Air Force assets et cetera?

In other words, the ship might not have to be tethered to a narrowly defined task force but you just need to be able to have the information that everybody else does so that you can make tactical or operational decisions to employ that ship to the max extent practical of its capabilities.

There is a significant shift underway. The question we are now posing is: "What capability do I need, and can I get it from a sister service that already has something that provides the weapons, the C2 or the ISR that I need?

I need to know how to exploit information which benefits either my situational awareness, my offensive or defensive capability of my organic force. But you don't necessarily need to own it in order to benefit from it.

And I think that when we really start talking about integration, that's probably one of the things that we could realize very quickly is that there are certain assets and data streams that come from the Air Force or the Navy that make the USMC a more lethal and effective force, and vice versa.

The key question becomes: "How do I get the most decisive information into an LHA/LHD? How do I get it into a marine unit so that they can benefit from that information and then act more efficiently or lethally when required?"

Question: We first met when you were at Eglin where you were working the F-35 warfare system into the USMC.[2]

Now that the F-35s are becoming a fact of life for both the U.S. services and the allies in the Pacific, how can we best leverage that integrable capability?

Col Gillette: *The development is a significant one. It is not only a question of interoperability among the F-35 fleet, it is the ability to have common logistical and support in the region with your allies, flying the same aircraft with the same parts.*

And the big opportunity comes with regard to the information point I made earlier. We are in the early stages of exploiting what the F-35 force can provide in terms of information dominance in the Pacific, but the foundation has been laid.

And when we highlight the F-35 as the 21st century version of what the World War II Navy called the big blue blanket with the redundancy and the amount of information that could be utilized, it's pretty astonishing if you think about it.

The challenge is to work the best ways to sort through the information resident in the F-35 force and then how do you utilize it in an effective and efficient way for the joint force.

But the foundation is clearly there.

Question: The new focus is on the maritime battle, and this requires a shift in USMC training. How are you approaching that challenge?

Col Gillette: *So long as I've been in the Marine Corps and the way that it still currently is today, marine aviation exists to support the ground combat.*

That's why we exist. The idea that we travel light and that the aviation element within the MAGTAF provides or helps to provide the ground combat element with a significant capability is our legacy.

We are now taking that legacy and adapting it. We are taking the traditional combat engagement where you have battalions maneuvering and aviation supporting that ground element and we are moving it towards Sea Control, and Sea Denial missions.

We are reimagining the potential of what the infantry does. That doesn't mean that they do that exclusively because, although I think that our focus in the Marine Corps, as the Commandant said, is shifting towards the Pacific that doesn't relegate or negate the requirement to be ready to respond to all of the other things that the Marine Corps does.

It might be less of a focus, but I don't think that that negates our requirement to deal with a variety of core missions.

It's a question of working the balance in the training continuum. What does an infantry battalion train to?

Do they train to a more traditional battalion in the attack or in the defense and then how do I use my aviation assets to support either one of those types of operations?"

As opposed to, "I might have to take an island, a piece of territory that we're going to use a mobile base, secure it so that we can continue to push chess pieces forward in the Pacific, in the Sea Control, Sea Denial end-state."

Those are two very different kind of skill sets. If there's one thing that the Marine Corps is very good at it's being very versatile and being able to switch from one to the other on relatively short order.

But in order to do that, you have to have a very dedicated and well thought out training continuum so that people can do both well, because if you say that you can do it the expectation is that you can do it well.

We are shaping a new Marine Littoral Regiment (MLR) but we're still in the nascent stages of defining what are the critical tasks that something like that needs to be able to do and then how you train to it.

How do we create not only the definition of the skill sets that we need to train large formations to, but then what venues must we have to train?

How to best combine simulated environments with real world training out on a range?

We're working through all that right now and it'll be interesting to watch how that process unfolds,

But it is definitely a mind shift to rethink the context in which our Ground Combat forces will conduct offensive of defensive operations, and specifically, what tasks they are expected to be capable of in this environment.

What we've done in the past is very well-defined and we have a very defined training continuum for those large formations. In this new role in the Pacific, that's something that I think over the next few years we'll get our arms around and we will learn from doing.

As we start to field these formations out to the Pacific we'll really start to figure out where are we good at training and where are gaps that we need to close and shape the venues and methods to fill those in those gaps.

We're constantly looking at new venues and new methods to start to do the things that we need to do with the new approach. For example, we are taking our TACAIR Community up to the Nellis range for large integrated strike missions.

We do face-to-face planning with the Air Force and Navy so that our students can understand the capabilities and limitations of these different platforms.

U.S. Marine Corps LtCol Steve Gillette, VMFA-121 squadron commander, operations and nuclear integration, explains the capabilities of the F-35B Lightning II joint strike fighter to Dr. Ng Eng Hen, Singapore minister of defense, at Luke Air Force Base, Ariz., Dec. 10, 2013. U.S. Air Force photo by Senior Airman Jason Colbert

They rub elbows with the USAF and Navy operators and gain first-hand knowledge of the strengths and weaknesses of these different platforms.

Then we fly them all back home and then the next night we go out with this huge armada of joint assets.

And it's, out of the assets that play on this, it's probably 50% Marines and the other 50% are Growlers, Air Force platforms et cetera. And then we do a mass debrief.

And this starts to chip away at the legacy perspective: "Okay, I'm a master of my machine." They come to WTI and learn how to think an integrated manner.

But more importantly, they get exposed and actually go out and do the integration with joint service assets to see the strengths and weaknesses so that they understand the planning considerations required for the joint fight against peer competitors and how to work beyond what their Marine Corps platform can do.

Another example is when we do what we call our Offensive Anti-Air Warfare (OAAW) Evolution. We fight peer versus peer against one another. We have real-time intelligent collects on what the other side is doing, so the plans change real-time, airborne and on the ground. There's deception; there's decoys.

It's pretty amazing to watch and oh, by the way, they get to use their weapons systems, their command-and-control systems to the fullest extent of their capabilities on both sides.

This allows us to engage a thinking, breathing enemy who is well-trained and has all the latest and greatest systems, but they do that with assets that not are resident just to the Marine Corps.

We operate with assets that come from the Department of Defense to show them the importance, on both sides, whether it's the C3 with their surface-to-air missiles and their own red fighters or the blue fighters with both organic assets, as well as national assets.

We are focused on operating, not just with the assets that you control, the ones that sit out on our flight line or sit in our command and control, but how these other things can contribute in the joint fight. And to shape effect methods to get the enabling information, digest it and then use it in near real-time.

It's pretty interesting to watch and the outcomes of this evolution are wildly different, based on the ability of the students to use these things that they're not used to working with, incorporate them in real-time into their plan and then execute.

Which I think, if you were to look at any high-end conflict or contingency where you have similarly matched forces in terms of training and gear.

That will make the difference between somebody who is wildly successful or *wildly unsuccessful, with your ability to direct and use those things real-time being a crucial delineator to combat success.*

Question: How do you see the growth of simulation in this training approach?

Col Gillette: *You can never just say, "I'm going to train only in a simulated environment." The simulated environment is good for a number of reasons. One key contribution is your ability to connect simulators, pick whatever platform it is.*

We are working with the surface warfare elements of the USN to incorporate synthetic/real training. What that will enable us to do is, take live fly events with their simulator event and start to fuse those two worlds, the simulated world as well as the live fly.

And this allows us to create, not only a complex, robust environment that has airplanes, real airplanes, synthetic airplanes, synthetic ships, both good and bad, but then go out to try and then solve a problem in that environment.

We're just starting to dip our big toe into this new environment, but what I think what we will find is that a surface warfare officer can learn how might a F-35 sense something that they would then prioritize high enough that they would want to shoot with one of their organic weapons.

If I could, I'd have every joint asset come to our WTI exercise, every class and integrate with our people. The reality is, due to real-world realties, these high-demand, low-density assets will not be free to come.

However, if I could create a simulated environment where I could get reps from an F-35 perspective, from a Viper, it doesn't matter what platform it is, but they get used to thinking about receiving and then executing off the information that would come from one of those high-demand, low-density assets.

I think what it will do is make our ability to then plug and play in a future contingency.

Another piece of the puzzle is to determine: how do we go from the simulator to a blending of live event with some amount of simulation mixed in there to create the contested environment?

And a lot of people define what is a contested environment differently, but what you'll be able to do is to create an environment which you actually go fly in, from Marine Aviation's perspective, against a threat that's both real and simulated.

We will shape a blended training environment as opposed to, "I do simulators and then I try to replicate it as best I can out on the range with real things."

There'll be requirements to have real things out on the range but there will be a blending, which, from the operator's perspective, will be no different than a completely live environment.

Working Mobile Basing

May 31, 2020

The USMC has mobile basing in its DNA. With the strategic shift from the Middle Eastern land wars to full spectrum crisis management, an ability to distribute a force but to do so with capabilities which allow it to be integrable is crucial.

For the Marines, this means an ability to operate an integrable force from seabases, forward operating bases (FOBs) or forward arming and refueling points (FARPs).

As the Marines look forward to the decade ahead, they are likely to enhance their capabilities to provide for mobile bases which can empower the joint and coalition force by functioning as a chess piece on the kill web enabled chessboard.

But what is required to do mobile basing?

What are the baseline requirements to be successful?

In a recent discussion with LtCol Barron, ADT&E Department Head at MAWTS-1, we had a wide-ranging discussion with regard to the flexible basing dynamic, and I will highlight a number of takeaways from that discussion.

ADT&E is focused on the core task of fighting today with the current force but also looking forward to how to enhance that force's capabilities in the near to mid-term as well.

Rather than quoting the LtCol directly, I am drawing on our discussion and highlighting what from my perspective are key elements for shaping a way ahead with what I call mobile basing,

The discussion with Lt. Col. Barron highlighted six key takeaways.

- The first one is the crucial need for decision makers to determine why a mobile base is being generated and what the tactical or strategic purpose of doing is. It takes time and effort to create a mobile base, and the mobile base commander will need to operate with mission command with regard to his base to determine how best to operate and for what purpose.
- The second one is the importance of determining the projected duration of the particular base. This will have a significant impact in shaping the question of logistics support. What is needed? How to get it there? And from what supply depot, afloat or ashore in adjacent areas?
- The third one is clearly the question of inserting the force into the mobile base and ensuring its optimal capabilities for survivability. What needs to be at the base to provide for organic survivability? What cross links via C2 and ISR will provide for an extended kill web to support the base and its survivability?
- A fourth one is to determine what the base needs to do to contribute to the wider joint or coalition force. With the evolution of technology, it is possible now to have processing power, and strike capabilities distributed and operated by a smaller logistics footprint force, but how best to configure that base to provide the desired combat effect for the joint or coalition force?
- A fifth one is clearly a crucial one for operating in a contested environment. Here the need is for signature control, or an ability to have as small a signature footprint as possible commensurate with achieving the desired combat effect. Signature management could be seen as a component of survivability. However, the management of signatures down to the small unit level requires a disruptive shift in our mindset.
- The sixth one is clearly having an exit strategy in mind. For how long should the force be at the mobile base? For

what purposes? And what needs to be achieved to enable the decision to move from the mobile base?

In effect, the discussion highlighted what one might refer to as the "three Ss": an insertion force operating from a variety of mobile bases needs to be able to be sustainable, survivable, and signature manageable.

With regard to current USMC capabilities, the MV-22, the C-130, the Viper, the Venom, the CH-53E and the F-35 are key platforms which allow the Marines to integrate and move a lethal combat force to a mobile base.

But the C2/ISR enablement is a key part of the requirement and the digital interoperability efforts are a key part of shaping a more effective way ahead. And in the relatively near term, the Ch-53K replacing the E is a key enabler for an enhanced mobile basing strategy.

It is clear that as the U.S. services work their way ahead in the evolving strategic environment, the USMC's core skill set with mobile basing will figure more prominently, and become a key part of the Marines working with the joint and coalition force in shaping a more effective way ahead for the integrated distributed force.

Moving Forward with Mobile Basing

June 3, 2020

As noted earlier, a key contribution which the USMC can provide for the joint and coalition force afloat or ashore is mobile and expeditionary basing.

As the joint and coalition force shapes greater capabilities through C2/ISR innovations and integrability of the sensor-strike kill web, that capabilities will be enhanced to operate distributed expeditionary basing for the insertion forces.[3]

But one fights with the force one has and builds forward from there. So where are the Marines currently with regard to mobile basing capabilities?

MAWTS-1

In a discussion with Major Brian Hansell, MAWTS-1 F-35 Division Head, it is clear that the coming of the F-35 to the USMC has expanded their ability to operate within a broader kill web and to both empower their expeditionary bases as well as to contribute to the broader kill web approach.

The Marine's F-35s are part of the broader joint and coalition force of F-35s, and notably in the Pacific this extends the reach significantly of the Marine's F-35s and brings greater situational awareness as well as reach to other strike platforms to the force operating from an expeditionary base as well as enhancing the kill web reach for the joint or coalition force.

As Major Hansell put it: *By being an expeditionary, forward-based service, we're effectively extending the bounds of the kill web for the entire joint and coalition force.*

U.S. Marine Corps AH-1Z Vipers, assigned to Marine Aviation Weapons and Tactics Squadron One (MAWTS-1), fly to a forward arming and refueling point during Weapons and Tactics Instructor (WTI) course 1-21, at Stoval Airfield, Dateland, Arizona, Oct. 16, 2020. Credit: MAWTS-1. This was a 9-ship FARP operation.

The F-35 brings a unique capability to the Marine Corps as it works mobile basing but reworking the assault force more generally is a work in progress. The digital interoperability initiative is a crucial one as the assault assets will have integrability they do not currently have, such as the Viper attack helicopter getting Link-16.

And the heavy lift element, which is a bedrock capability for the insertion force, is older, not easily integrable, and is in diminishing numbers. The CH-53K which is to replace it will provide significant enhancements for an insertion force operating from afloat or ashore mobile bases but needs to be ramped up in numbers capable of raising the combat level of the current force.

In a discussion with Major James Everett, Head of the Assault Support Department at MAWTS-1, we discussed the force that we have and some ways ahead for enhanced capability in the near to mid-term. The Assault Support Department includes several key divisions: CH-53, MV-22, KC-130, UH-1, and AH-1.

I had a number of takeaways from that conversation, and am not quoting Major Everett directly, as I am highlighting some key elements from our discussion but am also adding my own judgments with regard to what those key elements mean going forward.

- The first point is that indeed we need to focus on the force we have now, because we will fight with the force we have now. The Marines by being in the land wars for the past twenty years, have become part of the joint force, and have relied on elements from the joint force, that they would not necessarily have access to when doing force insertion in the Pacific. This means that the digital interoperability effort under way within Marine Corps innovation is not just a nice to have effort, but a crucial one to ensure that the insertion force package can work more effectively together and to leverage other key support assets which might be available from the joint or coalition force. After all, a mobile base is being put on the chessboard for a strategic or tactical objective and survivability is a key requirement.

- The second point is about sustainability. And sustainability is a function of the lift assets which can bring the kit and supplies needed for the duration of the mission. For the Marines, this is defined by KC-130J, CH-53E, MV-22, and UH-1Y lift support. And it is also defined by air refuelable assets to the assault force as well. The Marines have limited indigenous assets to provide aerial refueling which, dependent on the mission and the time scale of the force insertion effort, might need to depend on the Navy or Air Force for this capability.
- The third point is about C2. With the shift from the land wars, where the Marines were embedded within CENTCOM forces, C2 was very hierarchical. Working mission command for a force operating in a degraded environment is a key challenge, but one which will have to be met to deliver the kind of distributed mobile based force which the Marines can provide for the joint and coalition force, and not just only in the Pacific, but would certainly provide a significant capability as well for the fourth battle of the Atlantic.
- The fourth point is the clear importance of the coming of the CH-53K to the force. It is not only a question of a modern lift asset with significantly enhanced capabilities to provide for assault support, it is that it is a digital aircraft which can fully participate in an integrated distributed mission.
- The fifth point is that the digital interoperable initiative will not only provide for ways to better integrate assets but will enhance what those assets can do. A key example is the nature of what a Viper assault asset can do afloat as well ashore when operating with Link-16 and full motion video.
- The sixth point is that the coming of autonomous systems whether air or maritime can expand the

situational awareness of the insertion force, as long as signatures can be managed effectively. And for the insertion force this can be about autonomous systems transported to a base, operating from an afloat asset, or tapping into various overhead assets, such as Triton or Reaper. Or put another way, as digital interoperability is worked there will be expanded effort to find ways to support the insertion force operating from a mobile base. This will be an interactive process between what C2/ISR assets are available in the kill web, and how the Marines ashore or afloat can best use those resources. We have seen such a migration with the U.S. Navy as the CSG and fleet is adding MISR or Maritime ISR officers, and this change actually was inspired by the operations of 3rd MEF in Afghanistan. What we might envisage is simply the next iteration of what was done ashore with now the afloat and insertion forces in the maritime environment.

- The seventh point is the key emphasis on timeliness for a mobile basing option. It is about the insertion force operating within the adversary's decision cycle and operating to get the desired combat effect prior to that adversary being successful in getting his combat result, namely, eliminating or degrading the insertion force. This is another way to understand the key significance of how C2/ISR is worked between the insertion force and the wider air-maritime force.

In short, the Marines will fight with the force they have; and as far as near-term modernization, ensuring that digital interoperability is built in and accelerated, full use of what an F-35 wolfpack can bring to the insertion force, and the continuing modernization of the assault force staring with the coming of the CH-53K in sufficient numbers, these are all key ways ahead.

The Role of Heavy Lift

September 21, 2020

During my discussions earlier this year with several MAWTS-1 officers, we focused on the thinking and training of the USMC to further enhance their capabilities for mobile and expeditionary basing.

Heavy lift is a key capability to do so. And heavy lift really comes in two forms: fixed wing aircraft, and rotorcraft. My guide in the discussion of the lift-basing dynamic was Major James Everett, head of the Assault Support Department at MAWTS-1. As Major Everett put it: *A key focus of effort for the assault support community is upon how we can best assist through expeditionary basing to provide for sea control. We're trying to get away from any permanent type of land basing in a maritime contested environment.*

A key enabler for flexible basing inserts or operations from the maritime fleet, inclusive of the amphibious ships, are the capabilities which the Marine Corps has with its tiltrotor and rotorcraft community. This community provides an ability to insert a sizable force without the need for airstrips of the size which a KC-130J would need.

Or put in another way, the Marines can look at basing options and sustainability via air either in terms of basing options where a fixed wing aircraft must operate, or, in a much wider set of cases, where vertical lift assets can operate.

This could by sea, which depends on support by an amphibious or a Military Sealift Command (MSC) ship, but the challenge for the Marines is that moving bases deeper into the maritime area of operations creates enhanced challenges for the MSC and raises questions about viable sustainable options.

We have already seen this challenge with regard to the littoral combat ship fleet, where the MSC is not eager to move into the littorals to supply a smaller ship, but it is much more willing to take its ships into a task force environment with significant maritime strike capability to give it protection.

The most flexibility for the mobile or expeditionary basing options clearly comes from vertical lift support aircraft. The challenge is that the current CH-53E fleet has been heavily tasked by the more than a decade of significant engagement in the Middle East. The Marines unlike the U.S. Army did not bring back their heavy lift helicopters for deep maintenance but focused on remaining engaged in the fight by doing the just enough maintenance to continue safe and effective flight operations in theater.

As Major Everett put it: *The Army brought their helicopters back from Afghanistan and they'd strip them down to the frame and they'd rebuild them basically. We just didn't do that.*

This means that the heavy lift operational force inventory is relatively low compared to the required capabilities.

And as the focus shifts to the Pacific, with its tyranny of distance and the brutal operating conditions often seen in the maritime domain, having a very robust airlift fleet becomes not a nice to have, but a foundational element. The replacement for the CH-53E, the CH-53K, will provide a significant enhancement to the lift capability, and sustainability in operations as well.

It is also a question of being able to deliver combat support speed or CSS to the mobile or expeditionary base, and clearly the combination of tiltrotor and heavy lift can do that.

But the challenge will be having adequate numbers of such assets, notably, because the nature of the environment is very challenging, and the operational demand will go up significantly if one wants to operate a distributed force but one which is sustained and protected by an integrated force.

As Major Everett put it: *There's no way with the types of shipping and numbers of shipping we have, that we could possibly carry enough aircraft on that shipping to enable any type of land control without 53s.*

An aspect that makes the upgraded heavy lift fleet a key enabler for expeditionary basing will be the installation of a mesh network manager into the digital cockpit of the CH-53K, and its build into the legacy aircraft as well. This makes it part of an integrable force, not just an island presence force.

According to Major Everett: *The core kind of skills that 53 pilots train to, are not going to change. But obviously the physicality of the new helicopter is very different.*

It can lift more relevant materials or assets and in larger numbers. It holds the 463L pallets that allow for rapid off and on-loads from the fixed wing aircraft which could provide distribution points for the heavy lift fleet.

Additionally, the impact of the CH-53K's integrated digital interoperability and its integration into the kill web will be significant.

U.S. Marines assigned to Marine Aviation Weapons and Tactics Squadron One (MAWTS-1), prepare to offload cargo from a CH-53E Super Stallion onto a Light Capability Rough Terrain Forklift, during Assault Support Tactics 3 (AST-3) in support of Weapons and Tactics Instructor (WTI) course 1-21, at Kiwanis Park in Yuma, Arizona, Oct. 16, 2020. U.S. Marine Corps photo by Cpl. KarlHendrix Aliten.

In short, the desire to have a Marine Corps enhanced role in sea control and sea denial with an island strategy really enhances the importance of heavy lift helicopters, as demonstrated in the work ongoing at the WTIs at MAWTS-1.

Forward Arming and Refueling Points (FARPs)

June 8, 2020

When considering contributions which the USMC can make to the joint or coalition force in Pacific operations, an ability to put FARP on virtually any spot on the kill web chessboard is clearly a key contribution.

These are referred to as Forward Arming and Refueling Points (FARPs) but are really Arming and Refueling Points because where one might put them on the chessboard depends on how one wants to support the task forces within a kill web.

U.S. Marines with Marine Wing Support Squadron (MWSS) 371, Marine Wing Support Group (MWSG) 37, 3rd Marine Aircraft Wing (MAW), operate a tactical aviation ground refueling system (TAGRS) during a forward area refueling point (FARP) operation at Marine Corps Air Station Yuma, Feb. 4, 2020. The TAGRS enables the MWSS to rapidly establish a high-throughput, dual-point refueling site while maintaining critical mobility in austere locations making it a valuable asset for the MAW. U.S. Marine Corps Photo by Lance Cpl. Julian Elliott-Drouin.

In looking at a theater of operations, and certainly one with the tyranny of distance of the Pacific, one needs to be able to have a layer of fuel support for operations. For the Marines operating from the sea, this clearly includes combat ships, MSC tankers and related ships, as well as airborne tanker assets. By deploying a relatively small logistics footprint FARP or ARP, one can provide a much wider of points to provide fuel for the combat force. And in Marine terms, the size of that footprint will depend on whether that FARP is enabled by KC-130J support or by CH-53E support, with both air assets requiring significantly different basing to work the FARP.

I had a chance recently to discuss FARP operations and ways to rework those operations going forward with Maj Steve Bancroft, Aviation Ground Support (AGS) Department Head, MAWTS-1, MCAS Yuma.

There were a number of takeaways from that conversation which provide an understanding of how the Marines are working their way ahead currently with regard to the FARP contribution to distributed operations.

- The first takeaway is that when one is referring to a FARP, it is about an ability to provide a node which can refuel and rearm aircraft. But it is more than that. It is about providing capability for crew rest, resupply and repair to some extent.
- The second takeaway is that the concept remains the same but the tools to do the concept are changing. Clearly, one example is the nature of the fuel containers being used. In the land wars, the basic fuel supply was being carried by a fuel truck to the FARP location. Obviously, that is not a solution for Pacific operations. What is being worked on now at MAWTS-1 is a much more mobile solution set. Currently, they are working with a system whose provenance goes back to the 1950s and is a helicopter expeditionary refueling system or HERS system. This legacy kit limits mobility as it is very heavy and requires the use of several hoses and fuel separators. Obviously, this solution is too limiting so they are working a new solution set. They are testing a mobile refueling asset called Tactical Aviation Ground Refueling System (TAGRS).

- The third takeaway was that even with a more mobile and agile pumping solution, there remains the basic challenge of the weight of fuel as a commodity. A gallon of gas is about 6.7 pounds and when aggregating enough fuel at a FARP, the challenge is how to get adequate supplies to a FARP for its mission to be successful. To speed up the process, the Marines are experimenting with more disposable supply containers to provide for enhanced speed of movement among FARPs within an extended battlespace. They have used helos and KC-130Js to drop pallets of fuel as one solution to this problem. The effort to speed up the creation and withdrawal from FARPs is a task being worked by the Marines at MAWTS-1 as well. In effect, they are working a more disciplined cycle of arrival and departure from FARPs. And the Marines are exercising ways to bring in a FARP support team in a single aircraft to further the logistical footprint and to provide for more rapid engagement and disengagement as well.
- The fourth takeaway is that innovative delivery solutions can be worked going forward. When I met with Col. Perrin at Pax River, we discussed how the CH-53K as a smart aircraft could manage airborne MULES to support resupply to a mobile base. As Col. Perrin noted in our conversation: "The USMC has done many studies of distributed operations and throughout the analyses it is clear that heavy lift is an essential piece of the ability to do such operations."[4] And not just any heavy lift – but heavy lift built around a digital architecture.

Clearly, the CH-53E being more than 30 years old is not built in such a manner; but the CH-53K is. What this means is that the CH-53K "can operate and fight on the digital battlefield."

And because the flight crew are enabled by the digital systems onboard, they can focus on the mission rather than focusing primarily on the mechanics of flying the aircraft. This will be crucial as the Marines shift to using unmanned systems more broadly than they do now.

For example, it is clearly a conceivable future that CH-53Ks would be flying a heavy lift operation with unmanned "mules" accompanying them. Such manned-unmanned operations require a lot of digital capability and bandwidth, a capability built into the CH-53K.

If one envisages the operational environment in distributed terms, this means that various types of sea bases, ranging from large deck carriers to various types of Maritime Sealift Command ships, along with expeditionary bases, or FARPs or FOBS, will need to be connected into a combined combat force.

To establish expeditionary bases, it is crucial to be able to set them up, operate and to leave such a base rapidly or in an expeditionary manner (sorry for the pun). This will be virtually impossible to do without heavy lift, and vertical heavy lift, specifically.

Put in other terms, the new strategic environment requires new operating concepts; and in those operating concepts, the CH-53K provides significant requisite capabilities. So why not the possibility of the CH-53K flying in with a couple of MULES which carried fuel containers; or perhaps building a vehicle which could come off of the cargo area of the CH-53K and move on the operational area and be linked up with TAGRS?

I am not holding Maj. Bancroft responsible for this idea, but the broader point is that if distributed FARPs are an important contribution to the joint and coalition forces, then it will certainly be the case that autonomous systems will play a role in the evolution of the concept and provide some of those new tools which Maj. Bancroft highlighted.

MAWTS-1 and the F-35

May 29, 2020

Over the past few weeks, I have been discussing with USAF and U.S. Navy officers on how the two services are training to shape greater synergy with regard to the integrated distributed force.

The fusing of multiple sensors via a common interactive self-healing web enhances the ability of the entire force, including key partners and allies, to engage cooperatively enemy targets in a time of conflict.

Interactive webs can be used for a wide range of purposes throughout the spectrum of conflict and are a key foundation for full spectrum crisis management. To play their critical role when it comes to strike, whether kinetic or non-kinetic, this final layer of the web needs to have the highest standards of protection possible.

From the USAF and U.S. Navy perspectives, where does the USMC fit into the evolving kill web approach?

Clearly, one answer which has been given several times can be expressed in terms of one of the Marines key competences: bringing an integrated force to a mobile operational setting whether afloat or ashore.

It is important to consider a base afloat or ashore as part of the chessboard from a basing point of view. Too often when one mentions basing, the mind goes quickly to a fixed air or ground base, but in the evolving strategic environment, an ability to work across a wide variety of basing options is crucial.

And no force in the world is more focused on how to do this than the USMC. With the arrival of the USS America class LHA, the amphibious fleet moves out from its greyhound bus role to being able to contribute fully to sea control in transit or in operations, thereby relieving the U.S. Navy large deck carriers from a primary protection role.

The capability of the F-35s to hunt as a pack and through its CNI system and data fusion capabilities, the pack can work as one. The integration of the F-35 into the Marine Corps and its ability to work with joint and coalition F-35s provides significant reach to F-35 empowered mobile bases afloat or ashore

When I last visited MAWTS-1 at MCAS Yuma, I had a chance to discuss the evolving focus on mobile basing and learned that indeed the U.S. Navy and USAF were visitors to Yuma to discuss mobile basing.

Because the approach to mobile basing is being worked in the context of preparing for conflicts against full spectrum capable adversaries, in effect, the mobile basing approach will be about moving pieces on the kill web chessboard.

Recently, I had a chance to talk with Major Brian "Flubes" Hansell, MAWTS-1 F-35 Division Head, and I had a number of takeaways from my conversation with with him.

- The first takeaway is that following a significant focus on the land wars in the past twenty years combined with the return to the sea, the Marines are shaping new capabilities to operate at sea and in a way that can have significant combat effects on the expanded battlespace. And they are doing so from expeditionary bases, afloat and ashore. According to Major Hansell: *The Marine Corps is a force committed to expeditionary operations. When it comes to F-35, we are focused on how best to operate the F-35 in the evolving expeditionary environment, and I think we are pushing the envelope more than other services and other partners in this regard. One of the reasons we are able to do this is because of our organizational culture. If you look at the history of the Marine Corps, that's what we do. We are an expeditionary, forward-leaning service that prides itself in flexibility and adaptability.*
- The second takeaway is that the coming of the F-35 to the USMC has expanded their ability to operate within a broader kill web and to both empower their expeditionary bases as well as to contribute to the broader kill web approach. The Marine's F-35s are part of the broader joint and coalition force of F-35s, and notably in the Pacific this extends the reach significantly of the Marine's F-35s and brings greater situational awareness as well as reach to other strike platforms to the force operating from an expeditionary base as well as enhancing the kill web reach for the joint or coalition force. As Major Hansell put it: *By being an expeditionary, forward-based service, we're effectively extending the bounds of the kill web for the entire joint and coalition force.*

- The third takeaway is how the concepts of operations might empower a kill web approach. The F-35 is not just another combat asset, but at the heart of empowering an expeditionary kill web-enabled and enabling force. As Major Hansell put it: *During every course, we are lucky to have one of the lead software design engineers for the F-35 come out as a guest lecturer to teach our students the intricacies of data fusion. During one of these lectures, a student asked the engineer to compare the design methodology of the F-35 Lightning II to that of the F-22 Raptor. I like this anecdote because it is really insightful into how the F-35 fights. To paraphrase, this engineer explained that "the F-22 was designed to be the most lethal single-ship air dominance fighter ever designed. Period. The F-35, however, was able to leverage that experience to create a multi-role fighter designed from its very inception to hunt as a pack.* Simply put, the F-35 does not tactically operate as a single aircraft. It hunts as a network-enabled, cooperative four-ship fighting a fused picture, and was designed to do so from the very beginning. As Major Hansell put it: *We hunt as a pack. Future upgrades may look to expand the size of the pack.* The hunt concept and the configuration of the wolfpack is important not just in terms of understanding how the wolfpack can empower the ground insertion force with a mobile kill web capability but also in terms of configuration of aircraft on the sea base working both sea control and support to what then becomes a land base insertion force.
- The fourth takeaway focuses on the reach not range point about the F-35 global enterprise. The F-35 wolfpack has reach through its unique C2 and data fusion links into the joint and coalition force F-35s with which it can link and work. And given the global enterprise, the coalition and joint partners are working seamlessly because of common TTP or Tactics, Techniques, and Procedures. As Major Hansell put it: *From the very beginning we write a tactics manual that is*

distributed to every country that buys the F-35. This means that if I need to integrate with a coalition F-35 partner, I know they understand how to employ this aircraft, because they're studying and practicing and training in the same manner that we are. And because we know how to integrate so well, we can distribute well in the extended battlespace as well. I'm completely integrated with the allied force into one seamless kill web via the F-35 as a global force enabler.

- The fifth takeaway is the evolving role of the amphibious task force in the sea control mission. With the changing capabilities of strategic adversaries, sea control cannot be assumed but must be established. With the coming of the F-35 to the amphibious force, the role of that force in sea control is expanding and when worked with large deck carriers can expand the capabilities of the afloat force's ability to establish and exercise sea control. With the coming of the USS America Class LHA, the large deck amphibious ship with its F-35s onboard is no longer a greyhound bus, but a significant contributor to sea control as well. As Major Hansell noted: *The LHA and LHD can plug and play into the sea control concept. It's absolutely something you would want if your mission is sea control. There is tremendous flexibility to either supplement the traditional Carrier Strike Group capability with that of an Expeditionary Strike Group, or even to combine an ESG alongside a CSG in order to mass combat capability into something like an expeditionary strike force. This provides the Navy-Marine Corps team with enhanced flexibility and lethality on the kill web chessboard.*

- The sixth takeaway is that MAWTS-1 overall and the F-35 part of MAWTS-1 are clearly focusing on the integrated distributed force and how the Marines can both leverage an overall joint and coalition force able to operate in such a manner as well as how the Marines can maximize their contribution to the integrated distributed force. According to Major Hansell, *the CO of MAWTS-1, Colonel Gillette has put a priority on how to integrate as best as we* can with the Navy, as well as the joint force. And for the F-35 period of instruction during all Weapons Schools, we focus a tremendous amount of effort on integrating with the joint force, more so than I ever did on a legacy platform. We really strive to make our graduates joint integrators, as well as naval integrators. And I give Colonel Gillette (the current CO of MAWtS-1) all the credit in the world for moving us to that mindset and pushing us to learn how to operate in the evolving expeditionary environment.

In short, the USMC provides a critical piece of the kill web puzzle, as the United States and its allies rework their warfighting and deterrence strategies to deal with peer competitors worldwide.

A F-35B Lightning II fighter aircraft from Marine Fighter Attack Squadron (VMFA) 211, embarked on the Royal Navy aircraft carrier HMS Queen Elizabeth (R08), lands on the flight deck of the forward-deployed amphibious assault ship USS America (LHA 6) during flight operations between and America and the Royal Navy. America, flagship of the America Expeditionary Strike Group along with the 31st Marine Expeditionary Unit. August 20, 2021. Credit: USS America.

It is clearly a work in progress but new platforms are coming to the Marines, such as the CH-53K which clearly can support more effectively than the legacy asset, mobile basing, as well as the digital interoperability approach, which I have highlighted in recent interviews, which make the Marines more effectively woven into the kill web approach as well.

The Evolving Amphibious Task Force

June 4, 2020

The Marines are moving from a significant focus on the land wars to a "return to the sea." It has involved shaping and understanding what an air-mobile force could do when able to operate at greater reach into littoral regions with a rapid insertion force. And one empowered by the Ospreys coupled with fifth generation capability.

Under the twin influence of these two assets, the new LHA Class, the USS America ships, was introduced and with it, significantly different capabilities for the amphibious force itself. As the U.S. Navy reworks how it is operating as a distributed maritime force, which is being reshaped around the capability to operate a kill web force, the question of how best to leverage and evolve the amphibious force is a key part of that transition itself.

A case in point is how the Viper attack aircraft can evolve its roles at sea with the addition of key elements being generated by the digital interoperability effort, as well as adding a new weapons capability to the Viper, namely, the replacement for the Hellfire missile by the JAGM.

What this means is that the Viper can be a key part of the defense of the fleet while embarked on a variety of ships operating either independently, or as part of an amphibious task force.

Because the Viper can land on and operate from of a wide range of ships, thus enabling operational and logistical flexibility, and with integration of Link 16 and full motion wave forms as part of digital interoperability improvements, the Viper can become a key member of the kill web force at sea.

In discussions with Major Thomas Duff and Mr. Michael Manifor, HQMC Aviation, APW-53, Attack and Utility Helicopter Coordinators, I learned of the evolving mission sets which Viper was capable of performing with the digital interoperability upgrades.

With the upgrades coming soon via the digital interoperability initiative, the Viper through its Link 16 upgrade along with its Full-Motion video access upgrade, can have access to a much wider situational awareness capability which obviously enhances both its organic targeting capability and its ability to work with a larger swath of integrated combat space.

This means that the Viper can broaden its ability to support other air platforms for an air-to-air mission set, or the ground combat commander, or in the maritime space….

Because it is fully marinized, it can land and refuel with virtually any ship operating in the fleet, which means it can contribute to sea control, which in my view, is a mission which the amphibious task force will engage in with the expanded reach of adversarial navies.[5]

U.S. Marines with the 31st Marine Expeditionary Unit refuel an AH-1Z Viper and a UH-1Y Venom during Forward Arming and Refueling Point operations on Okinawa, Japan, Feb. 1, 2022. U.S. Marines from the 3d Marine Division conducted force-on-force training with the 31st MEU, providing an opportunity to test both seizing and defending key maritime terrain against a peer-level opponent. The training demonstrates the joint force's ability to integrate command and control, expand battlefield awareness, and conduct long-range precision strikes as an enduring stand-in-force. U.S. Marine Corps photo by Cpl. Ujian Gosun.

Recently, I had a chance to discuss with Major "IKE" White the AH-1Z Division Head at MAWTS-1, the evolution of Viper enabled by upgrades for fleet operations as well as its well-established role in supporting the ground maneuver force. In that conversation, there were a number of takeaways which highlighted potential ways ahead.

- The first takeaway is that the Marine Corps' utility and attack helicopters have been part of integrated operations and escort tasks throughout the land wars and can bring that experience to bear in the return to the sea. The Viper and the Venom have provided airborne escorts for numerous Amphibious Ready Groups over the last decade, partnering with destroyers, MH-60 Sierra and MH-60 Romeo to protect amphibious warships as they transited contested waterways.
- The second takeaway is the coming of the JAGM, which will provide a significant strike capability for the maritime force in providing for both sea control and sea denial. This missile provides increased lethality through a dedicated maritime mode, enhanced moving target capability, and selectable fusing; providing capability against both fast attack craft and small surface combatants. Millimeter wave (MMW) guidance increases survivability by providing a true fire-and-forget capability, removing the requirement for a terminal laser. Coupled with the AIM-9 sidewinder, the Viper will be able to engage most threats to naval vessels. The Viper's flexibility will provide even the most lightly defended vessels with a complete air and surface defense capability.
- The third takeaway is that by working integration of the MH-60 Romeo helicopter with Viper, the fleet would gain a significant defense at sea capability. Integration of the two helicopters within the amphibious task force

would allow them to provide an integrated capability to screen and defend the flanks of the afloat force. The MH-60 crews are optimized to integrate into the Navy's command and control architecture, and with onboard sensors can help detect potential targets and direct Vipers to engage threats. The integration of Link-16 will make this effort even more seamless. My interviews with NAWDC have underscored how the Navy is working through the question of how the integrable air wing will change when the MQ-25 joins the fleet, and working ways for the Romeo to work with MQ-25 and Advanced Hawkeye will inform Romeo as part of its fleet defense function. Clearly, integrating Romeos which fly onboard the amphibious class ships with the Viper would provide a significant enhancement of the flank defense capabilities for the amphibious task force. Working a Romeo/Viper package would affect the evolution of the Romeos as well that would fly off of the L class ships as well. And all of this, frees up other surface elements to support other missions at sea, rather than having to focus on defending the amphibs as greyhound buses.

- The fourth takeaway is that clearly this new role would have to be accepted and trained for, but I would argue, that in general, the U.S. Navy needs to rethink how amphibious ships can operate in sea control and sea denial functions in any case. I would argue as well that the enhanced efforts at digital interoperability within the USMC aviation force needs to be accompanied by upgrades of the elements of the amphibious task force with regard to C2/ISR capabilities as well. We are seeing Maritime ISR or MISR officers placed within the Carrier Strike Groups but they could be proliferated more broadly within the fleet.

In short, the evolution of the Viper with digital interoperability and with a new weapons package can clearly contribute to the evolution of the amphibious task force as it embraces sea control and sea denial missions and these missions will be crucial to supporting insertion forces moving to ashore expeditionary bases as well.

The Ground Combat Element in the Pacific Reset

September 24, 2020

As the USMC reworks its relationship with the U.S. Navy, a core focus is upon how the Marine Corps can provide for enhanced sea control and sea denial. A means to this end is an ability to move combat pieces on the chessboard of the extended battlespace.

But where does the ground combat element fight into this scheme for maneuver?

The key is to ensure that the USMC is combat capable today as it transitions to a new GCE that is lighter and more capable of tapping not into the air-maritime joint force, above and beyond what USMC integration provides.

The GCE faces the challenge of dealing with more traditional tasks as well as adapting to the evolving reconfiguration for the maritime fight. And it is a major shift facing the GCE for sure. The GCE is shifting from its most recent experiences of fighting in the land wars as a primary mission to providing support to, in Major Fitzsimmons, the Ground Combat Department Head at MAWTS-1, words, *a more amphibious distributed force operation. And in my view, this is a very big shift.*

Major Fitzsimmons provided a very helpful entry point into this discussion by recalling the earlier work which the Marines had done with the Company Landing Teams.

As Major Fitzsimmons put it: *The Company Landing Team was an experiment at how do we lighten the footprint of the force while still giving them the capabilities of what we see in larger forces today.*

To do that, we would leverage digital interoperability, connectivity, and reach back to weapon systems, to information, to targeting, to any of those capabilities that you generally see at some of the higher echelons that were not organic to a infantry company at that time.

The challenge then is to ensure that the infantry company has access to those types of capabilities and mature the force.

U.S. Marines with 3d Marine Littoral Regiment, 3d Marine Division establish a combat operations center during exercise Bougainville II at Puuola Range, Hawaii, Oct. 28, 2022. To adapt to developing battlefield conditions, alpha command conducted battle turnover and displaced. Bougainville II is a field exercise that allows the MLR and its subordinate units to conduct expeditionary advanced base operations across the island of Oahu. BVII displays the MLR's ability to rapidly establish and displace expeditionary advanced bases while executing command and control in a dispersed environment. U.S. Marine Corps photo by Lance Cpl. Cody Purcell

What Major Fitzsimmons meant by maturing the force was discussed later in the conversation. He highlighted the importance of having Marines earlier in their career able to work with various elements of the joint force, because they would need to leverage those capabilities as part of the more distributed GCE.

The Company Landing Team experiment also raised questions about equipment and personnel relevant to the current focus on building Marine Littoral Regiments (MLR).

According to Major Fitzsimmons: *How do we reinforce the CLT and how do we augment it with enablers?*

How do we augment it or enhance it with more proficient and more experienced fires personnel?

How do we augment it with small UAS capabilities?

How do we augment and enhance it with digital interoperability?

How do they communicate with their organic radios across multiple waveforms?

Who are they talking to?

What is their left and right for decisions?

Do they have fires approval?

Would the company commander have fires approval, or would he have to do what we were having to do in Afghanistan and Iraq, where I've got to call my boss and then the boss's boss, in order to get fires employed?

With the introduction of the new MLR, it is clear that these aspects of the CLT experiment are relevant to the way ahead.

Major Fitzsimmons is an infantry officer with fires experience at the company and battalion level, and clearly is focused on the key aspect of how you enable smaller and less organically capable forces in the extended battlespace and ensure that they have adequate fires to execute its missions.

And in dealing with peer competitors, clearly the ability to link the GCE with fires requires the right kinds of communication capabilities. As Major Fitzsimmons put it: *We are going to have to be significantly more distributed and quieter with respect to our emissions signatures than we have in the past.*

A major challenge facing the GCE is the range of adaptability that they will have to be able to deliver and operate with in the future. As Major Fitzsimmons put it: *I think the biggest shock to my community is going to be the level of adaptability that we're going to have to be able to achieve. We are going to have to train smaller forces to operate more autonomously and to possess the ability to achieve effects on the battlefield previously created at higher echelons.*

He focused as well on the tailorable aspect envisaged as well. *We will need to be tactically tailored to achieve whatever effect we need. It should be* akin to a menu; based on the mission and the effects needed to shape the environment towards mission accomplishment, we will need this capability or that capability which may require each element to be manned and equipped differently.

Then there is the challenge of the sustainability of the tailored force. How to ensure the logistics support for the distributed maritime focused USMC GCE?

Blue Water Expeditionary Operations and the Challenges for Aviation Ground Support Element

September 27, 2020

A key element for an evolving combat architecture clearly is an ability to shape rapidly insertable infrastructure to support Marine air as it provides cover and support to the Marine Corps ground combat element. This clearly can be seen in the reworking of the approach of the Aviation Ground Support (AGS) within MAWTS-1 to training for the execution of the Forward Air Refueling Point mission.

During my visit to MAWTS-1 in early September 2020, I had a chance to continue an earlier discussion with Maj Steve Bancroft, Aviation Ground Support (AGS) Department Head, MAWTS-1, MCAS Yuma.

In this discussion it was very clear that the rethinking of how to do FARPs was part of a much broader shift in in combat architecture designed to enable the USMC to contribute more effectively to blue water expeditionary operations.

The focus is not just on establishing FARPs, but to do them more rapidly, and to move them around the chess board of a blue water expeditionary space more rapidly. FARPs become not simply mobile assets, but chess pieces on a dynamic air-sea-ground expeditionary battlespace in the maritime environment.

Given this shift, Major Bancroft made the case that the AGS capability should become the seventh key function of USMC Aviation.

Currently, the six key functions of USMC Aviation are: offensive air support, anti-air warfare, assault support, air reconnaissance, electronic warfare, and control of aircraft and missiles. Bancroft argued that the Marine Corps capability to provide for expeditionary basing was a core competence which the Marines brought to the joint force and that its value was going up as the other services recognized the importance of basing flexibility,

But even though a key contribution, AGS was still too much of a pick-up effort. AGS consists of 78 MOSs or Military Operational Specialties which means that when these Marines come to MAWTS-1 for a WTI, that they come together to work how to deliver the FARP capability.

As Major Bancroft highlighted: *The Marine Wing Support Squadron is the broadest unit in the Marine Corps. When the students come to WTI, they will know a portion of aviation ground support, so the vast majority are coming and learning brand new skill sets, which they did not know that the Marine Corps has. They come to learn new functions and new skill sets.*

His point was rather clear: if the Marines are going to emphasize mobile and expeditionary basing, and to do so in new ways, it would be important to change this approach. Major Bancroft added: *I think aviation ground support, specifically FARP-ing, is one of the most unique functions the Marine Corps can provide to the broader military.*

He underscored how he thought this skill set was becoming more important as well. *With regard to expeditionary basing, we need to have speed, accuracy and professionalism to deliver the kind of basing in support for the Naval task force afloat or ashore.*

With the USMC developing the combat architecture for expeditionary base operations, distributed maritime operations, littoral operations in a contested environment and distributed takeoff-vertical landing operations, reworking how to execute FARP operations is a key aspect. FARPs in the evolving combat architecture need to be rapidly deployable, highly mobile, maintain a small footprint and emit at a low signature.

While being able to operate independently they need to be capable of responding to dynamic tasking within a naval campaign. They need to be configured and operate within an integrated distributed force which means that the C2 side of all of this is a major challenge to ensure it can operate in a low signature environment but reach back to capabilities which the FARP can support and be enabled by.

A U.S. Navy MH-60S Seahawk with Helicopter Maritime Strike Squadron (HSM) 51 takes off at a forward arming and refueling point (FARP) during Stand-in Force Exercise on Kin Blue, Okinawa, Japan, Dec. 11, 2022. SiF-EX is a Division-level exercise involving all elements of the Marine Air-Ground Task Force focused on strengthening multi-domain awareness, maneuver, and fires across a distributed maritime environment. U.S. Marine Corps photo by Lance Cpl. Emily Weiss.

This means that one is shaping a spectrum of FARP capability as well, ranging from light to medium to heavy in terms of capability to support and be supported. At the low end or light end of the scale one would create an air point, which is an expeditionary base expected to operate for up to 72 hours at that air point. If the decision is made to keep that FARP there longer, an augmentation force would be provided and that would then become an air site.

Underlying the entire capability to provide for a FARP clearly is airlift, which means that the Ospreys, the Venoms, the CH-53s and the KC130Js provide a key thread through delivering FARPs to enable expeditionary basing.

This is why the question of airlift becomes a key one for the new combat architecture as well. And as well, reimagining how to use the amphibious fleet as lilly pads in blue water operations is a key part of this effort as well.

In effect, an ability to project FARPs throughout the blue water and littoral combat space supporting the integrated distributed force is a key way ahead and is being worked at MAWTS-1.

The F-35 and USMC-US Navy Integration

September 29, 2020

During my visit to MAWTS-1, I met with Major Shockley, an F-35 instructor pilot at MAWTS-1, whose most recent F-35 experience has been in the Pacific with the squadron in Japan. He reinforced Col. Gillette's point in terms of the ability of USMC F-35s to work with allied, USAF and U.S. Navy F-35s as well to shape a situational awareness and strike force which expanded the reach of the joint or coalition force.

Indeed, Major Shockley highlighted the impact of F35-B thinking on base mobility. The F-35As and F-35Cs have some advantages in terms of fuel, and then range and loitering time with regard to the B, notably with regard to the C.

Because the force is so inherently integrable, how best to work the chessboard of conflict with regard to where the various F-35 pieces move on the chessboard. From this standpoint, he argued for the importance of shaping a "rolodex of basing locations" where F-35s could land and operate in a crisis.

He had in mind, not only what the very basing flexible B could provide but thinking through deployment of "expeditionary landing gear" to allow the As and Cs to operate over a wider range of temporary air bases as well.

Here he was referring to preparing locations with the gear to enable landing on shorter run "airfields" as well as the kind of modifications the Norwegians have done with their F-35s enabling them to land in winter conditions in the High North as well.

With the F-35B as well, a much wider range of afloat assets are being used to enable the F-35 as a "flying combat system" to operate and enable ISR, C2 and strike capabilities for the joint and coalition force. This is being demonstrated throughout the amphibious fleet, a fleet which can be refocused on sea control and sea denial rather than simply transporting force to the littorals.

U.S. Marines with Marine Fighter Attack Squadron 314 and Marine Aerial Refueler Transport Squadron 352, Marine Aircraft Group 11, 3rd Marine Aircraft Wing, conduct a new expeditionary landing demonstration with M-31 arresting gear Interim Flight Clearance (IFC), on Marine Corps Air Ground Combat Center Twentynine Palms, Calif., Dec. 3rd, 2020. This new capability allows the F-35C Lightning II to land on smaller runways anywhere in the world and ensures extended flexibility in combat operations. U.S. Marine Corps photo by Cpl. Cervantes, Leilani

A key consideration when highlighting what the F-35 as a wolfpack can bring to the force is deploying in the force multiples that make sense for the force. This rests upon how the combat systems are configured on that force. In simple terms, the integrated communications, navigation and identification systems operate through a multiple layer security system, allowing a four ship F-35 force to operate as one. And of course, Marines fly both the F-35B and the F-35C, the later when operating from large deck carriers.

With the Block IV software coming into the fleet, now an eight ship F-35 force can operate similarly. This allows for wolfpack operations and with the ability of the reach of the F-35 into other joint or coalition F-35 force packages the data flowing into the F-35 and the C2 going out has a very significant reach and combat impact.

This is not widely known or understood but provides a significant driver of change to being able to operate and prevail in denied combat environments. Leveraging this capability is critical for combat success for the U.S. and allied forces in the Pacific.

Unmanned Air Systems and the USMC

October 22, 2020

The primary use for UASs in the USMC has been in terms of ISR in the land wars, but with the return to the sea and now the focus on how the Marines can best help the U.S. Navy in the maritime fight, the focus has shifted to how to best use UASs in the maritime domain.

With the recent decision to cancel its MUX ship-based UAS to pursue a family of systems, the focus will be upon both land-based and sea-based UASs but not to combine these capabilities into a single air vehicle.

But the path to do this is not an easy one. And it is a path which is not just about the technology, but it is about having the skill sets to use whatever system is developed, the connectivity so that the combat effect can be connected to the maneuver force, and to have communication links which have low latency, notably in the maritime fight.

During my visit to MAWTS-1 in early September 2020, I had a chance to talk with Captain Dean, an experienced UAS officer who is a UAS instructor pilot at MAWTS-1.

We discussed a wide range of issues with regard to UAS within the USMC, but one comment he made really gets at the heart of the transition challenge: "What capabilities do we need to continue to bring to the future fight that we currently bring to the fight?"

What this question highlights is there is no combat pause for the Marines – they need to be successful in the current range of combat situations, and to re-shape those capabilities for the combat architecture re-design underway.

But what if this is not as significant and overlapping as one might wish?

This is notably true with regard to UAS systems. In general terms, the UAS systems which have been dominant in the Middle East land wars have required significant manning, lift capability to move them around in the battlespace and are not low-latency communications systems.

Although referred to as unmanned, they certainly are not so in terms of support, movement of exploitation systems, or how that data gets exploited.

There clearly is a UAS potential for the blue water and littoral engagement force but crafting very low demand support assets, with low latency communications are not here as of yet.

And in the current fights ashore, UASs, like the Blackjack, provide important ISR enablement to the Ground Combat Element. And as the Marines have done so, they have gained very useful combat experience and shaping of relevant skill sets to the way ahead for the UAS within the future force.

The goal is to have more flexible payloads for the UAS force going forward, but that means bringing into the UAS world, experienced operators in fields broader than ISR, such as electronic warfare.

But there is clearly a tension between funding and fielding of larger UAS's for the amphibious task force, and between shaping new systems useable by combat teams.

And the challenge here clearly is to manage information and to distribute by communications system. Although the phrase about distributing information at the right time and at the right place sounds good, this is very difficult to do, if the data links simply do not expose the combat force to adversary target identification.

This is yet another key area where contested combat space has not much to do with what can do with UASs in uncontested air space.

Captain Dean underscored that since 2015: *We have been able to normalize unmanned aviation with the USMC. We have been able to bring in a lot of experience into the VMUs and with the sundowning of the Prowlers, have brought in Marines experienced with electronic warfare as well. We continue to prioritize our training on the Blackjacks going to the MEUs.*

U.S. Marines with Marine Unmanned Aerial Vehicle Squadron (VMU) 1 launch an RQ-21 Blackjack in support of Weapons and Tactics Instructor course 2-18 at Cannon Air Defense Complex, Yuma, Ariz., March 22, 2018. Credit: MAWTS-1

He highlighted that this posed a challenge for transition. To get full value out of the Blackjacks operating off the amphibious force, changes need to be made on those ships to get full value from operating these UASs. But if the Blackjack is a short term or mid-term solution, the kind of investment which needs to be made is not likely to happen.

What he highlighted was the crucial importance of the infrastructure afloat to make best use of the UASs which the USMC and US Navy will operate. And given the challenge of managing space onboard the ship, sorting out the nature of the infrastructure and how to manage it is a key aspect of the way ahead for UASs.

Another challenge is who wants what within the combat force. If we are looking at the fleet as a whole, the desire is to have fleet wide ISR, or capabilities to deliver combat effect. If one is focused on the battalion, they are focused on having capabilities organic to the battalion itself.

Again, this is a development and investment challenges which as well raises questions of what kind of infrastructure can be developed to deal with each of these different operational level requirements.

Captain Dean asked: *What does the MAGTF want? What does the battalion want? These are not the same things.*

In short, a key question facing the Marines with regard to UASs: "What capabilities do we need to continue to bring to the future fight that we currently bring to the fight?"

7

2023 Interviews and Visit

In April of 2023, Laird interviewed the CO of MAWTS-1, Col Eric Purcell, and then visited the command in November after the second WTI of the year. This provided a chance to discuss how MAWTS-1 had progressed in working enhanced force mobility for the USMC within the broader joint force, a key emphasis of the force design effort.

The challenge is that while the Marines are working FARPs and other means to enhance force mobility, the joint force is in the throes of significant change, whether it be the U.S. Navy working distributed maritime operations or the USAF working agile combat employment.

How does the USMC effort to reorganize and enhance its contribution to the joint force while the joint force is itself in fundamental change with much uncertainty over how to do maritime distributed operations and the agile combat air combat employment?

The Navy and Air Force sides of this transition have been a major part of our work published elsewhere and provide insights with regard to how challenging the overall force transformation is within which the USMC is working to find its proper place.[1] It is not just up to MAWTS-1 to work the training for such an effort, but NAWDC and Nellis are clearly involved as well.

To put it simply: it is a work in progress and the Marines emphasis on a MAGTF organizing principle remains important going forward despite the effort to find ways to operate from much smaller organizational formations.

MAWTS-1 Works Mobile Basing and Support for the Distributed Joint Force

April 26, 2023

Ever since 2018, MAWTS-1 has focused on the high-end fight component of the full spectrum of warfare. Force distribution is a key part of the survivability against a competitor who has significant firepower and can concentrate fires on relatively fixed positions.

The Marines have worked mobile basing for a long time, such as working forward refueling points and buying the Osprey and the F-35B which can operate off a wide variety of launch and landing points.

But in the past few years, the emphasis has been about how to move more quickly from mobile operating bases and to do so in support of the joint force. This is a capability not only of interest to the Marines and the U.S. forces but core allies as well.

During Laird's latest visit to Australia, he highlighted the Australian Defence Forces's enhanced interest, for example, in agile air operations, and noted their interest to the CO of MAWTS-1, Colonel Eric Purcell. Purcell noted that given the close working relationship which the Marines had with the Australians, they were focused on training for such distributed force operations.

Col Purcell mentioned that last November his team met in London with U.S, and partner commands similar to MAWTS-1 in the UK. The USAF and the U.S. Navy along with Canada, and Australia discussed joint learning and training perspectives.

According to Purcell: *At the meeting last November, we looked at a number of different ways in which we can work jointly on problems such as agile combat employment, distributed maritime operations, Expeditionary Advanced*

Base Operations *(EABOs)*, and F-35 integration. Canada has just recently formally joined the F-35 program, so they were not part of that discussion.

As the Marines operate Ospreys. F-35s and now CH-53Ks, the Marines are bringing significantly capability to the evolving mobile basing function.

Mobile basing is playing a central role in the current phase of USMC transformation.

Col Purcell put it succinctly: *We are taking capability which we have had for some time, but focused on how we can move more rapidly from mobile base to mobile base. We have to find ways to make mobile bases, smaller, more distributed and persist for shorter periods of time.*

Another key aspect is that what has been a core competence of the USMC now is becoming a key capability for the wider joint and coalition force.

Col Purcell put it this way: *I think the challenge for all the forces, whether it's the Air Force, the Army, the Navy, the Marine Corps, or the coalition forces is that the sustainment of distributed forces is challenging. How do we adapt our maintenance, logistical and sustainment systems that have been used to operating from austere bases, but now enhance the mobility of those austere bases?*

During the 2020 visits, ground artillery Marines discussed the challenge of integrating their fires into a joint fires solution when emphasizing force distribution and mobility.

Col Purcell was asked about progress in this area. He argued that the joint fires piece is a central challenge being worked. He noted that at the recent WTI 2-23 they were working this hard. One example was incorporating the simulated integration of the future USMC Nemesis ground launch system into joint naval fires.

Part of the enhanced capability for the Marines to support force mobility was the involvement of four CH-53Ks into WTI 2-23.

Col Purcell indicated: *During the course we lifted 36K loads with the CH-53K which points to future capabilities. With regard to future capabilities, we can leverage the aircraft's ability to hold 9 to 10,000 pounds of fuel off on each of the three hooks of the CH-53K.*

The ability of each of the hooks to carry a fuel bladder is a key advantage for force mobility.

One could add that the changes in the cockpit allow for the management of such a load as well.

This is a real game changer for us at a time when we and the joint force are emphasizing distributed force logistical support and sustainability.

The Importance of Integration and Ownership in the Joint Fight: A Conversation with LtCol Barron at MAWTS-1

November 20, 2023

In Laird's discussions with MAWTS-1 in 2020, one of the most perceptive of the officers with whom he discussed the challenges facing force integration was LtCol Barron, ADT&E Department Head at MAWTS-1.

Earlier, in a 2018 interview with the then head of ADT&E, LtCol Schiller, he described a key function of ADT&E as assisting in the process of informing future requirements.

It is part of our mission to help requirement officers in Headquarters Marine Corps. We do this by taking items from DARPA, research labs, industry and the PMAs and integrate them into WTI courses. We then provide an after-action report with our assessment on their performance and utility to the force.

In other words, ADT&E is focused on the core task of fighting today with the current force but also looking forward to how to enhance that force's capabilities in the near to mid-term as well.

As Barron faces the end of his career with the USMC, Laird discussed with him a key challenge facing the U.S. forces, namely, not getting full value out of the systems which they already possess such as an F-35 global force.

He was asked how we could address this shortfall.

LtCol Barron: *I think you're absolutely right. The problem we face is how do we leverage these unique and disruptive capabilities that America and our coalition partners have because we're not getting the full benefit of them.*

And I think the way to maximize the use of the fantastic systems that we have is by further integrating our people.

I think intelligence, command and control, and fires are the elements that need further integration. And when we think about intelligence, it is, all of our collectors, whether it is a Fifth Gen aircraft that does a great job of sensing the environment, all the way down to individual threat sensors we need to be sharing that information to make a combined intelligence picture for our intelligence community, need to feed it through our command-and-control elements.

And I don't mean a single command and control element, just like everything's sensing, everything is contributing to command and control architecture, and then enabling the decision makers human on the loop or in the loop, depending on the situation to enable both kinetic and non-kinetic fires.

U.S. Marine Corps Maj. Paul R. Barron, right, a UH-1Y Venom instructor assigned to Marine Aviation Weapons and Tactics Squadron One (MAWTS-1), inspects cargo to be lifted during Weapons and Tactics Instructor Course (WTI) Course 2-15 near Yuma, Ariz., April 25, 2015. U.S. Marine Corps photo by Lance Cpl. Jodson B. Graves, 2nd MAW Combat Camera/Released.

That's very long answer.

But at the end of it, the thing that I think will enable this is ownership within each community. Ownership that this is our fight, that it's our responsibility. It's really easy to say, 'I'm an attack helicopter pilot, that's my job, I don't need to worry about it.' It's hard to say, I'm going to support digital interoperability by passing what I see through a command-and-control architecture to someone who can make a decision.

It's also my job to win. And the way I do that is feed the common operating picture for the decision maker.

We then addressed the question of building an operational

common operational picture and what that really means for a combat force.

LtCol Barron: *We say we want a Common Operating Picture (COP). What is the reality of that? Are we going to all have a single COP? Or are we going to have multiple systems that talk together?*

There are so many different communities. Within the Marine Corps, you've got aviation command and control, you've got the ground combat element that has its own systems, you have intelligence cops, and that is just within one service.

So how do we get everybody on the same picture? Or how do we share that information? It's, a struggle. That's where in the near term, I think the rubber hits the road. What's our common message format?

He then highlighted a community within the USMC where he thinks such progress is being made.

LtCol Barron: *I think one of the ways ahead is for the communities to want that interoperability. I have found the V 22 community is really on board with their mesh network manager in the back of their aircraft. That community has embraced that airborne gateway.*

And it's phenomenal. They are medium lift pilots and crews, but if you talk to any one of our students from that community you can have a great conversation about what's going on in the back of their aircraft with respect to waveform message formats and which antenna is doing which type of transmission. It's truly remarkable.

Ownership and improved integration are cost effective force multipliers that dwarf the capability of standalone new systems.

The Marine Corps Works the Next Phase of their Use of UAVs: The Perspective from MAWTS-1

November 21, 2023

Laird wrote a chapter in the 2018 book entitled, *One Nation Under Drones*, which focused on the experience of the USMC with UAVs to date. He wrote this piece as the Marines were shifting from the primary focus on the land wars and to an enhanced focus on amphibious operations. During operations in Iraq and Afghanistan, the Marines joined in with the U.S. Army and used the Shadow unmanned aerial system, for similar operations as the U.S. Army was engaged in the land wars.

But concurrently with the introduction of Shadow into the Corps, the ScanEagle was also introduced. And this system would fit the trajectory of the evolution of the Corps as it moved from a primary occupation with the land wars to a "return to the sea" and the joining of unmanned systems to the significant evolution of the Amphibious-Ready Group and Marine Expeditionary Unit pairing into a flexible amphibious ready task force, a change driven initially by the introduction of the Osprey but being reshaped as other manned aircraft systems come to the force and unmanned systems woven into the overall force insertion capability of the amphibious task force.

The Scan Eagle-Blackjack transition was part of the shift in focus from the land wars to amphibious at sea operations. When Laird wrote the essay the focus was upon shaping capabilities to be launched from a ship to support the ground maneuver element. At the time, the Marine Corps leadership was focused on a program called MUX (MAGTF Unmanned eXpeditionary UAS) which the aviation plan at the time projected initial operations in the 2025 time frame.

But as the then Deputy Commandant of Aviation, LtGen Rudder noted in 2020, that the MUX was being shelved in favor of a different approach. As was noted in a March 10, 2020 USNI News article:

I think what we discovered with the MUX program is that it's going to require a family of systems. The initial requirement had a long list of very critical requirements, but when we did the analysis and tried to fit it inside one air vehicle," they realized they had competing needs, Rudder said.

With a family of systems approach, my sense is we're going to have an air vehicle that can do some of the requirements, some of the higher-end requirements, potentially from a land-based high-endurance vehicle, but we're still going to maintain a shipboard capability, it just may not be as big as we originally configured."

The MUX program – formally the Marine Air-Ground Task Force (MAGTF) Unmanned Aerial System (UAS) Expeditionary – was meant to be a Group 5 UAS, the largest of the categories with highest altitude and greatest endurance. It would cover seven missions: command, control and communication; early warning; persistent fires; escort; electronic warfare; reconnaissance, intelligence, surveillance and target acquisition (RISTA); and tactical distribution...

Program officials realized they had a huge task ahead of them with so many separate missions, though, and early industry talks showed it may become cost-prohibitive. The seven missions were later sorted into two tiers of priority.

Still, as Rudder said, it became clear that those higher priority missions were incompatible with shipboard launch and recovery.

Power output and weight capacity, obviously you get more weight and power output with a ground-based system with a longer runway, expeditionary runway, than you can coming vertically off the back of a ship. Shipboard compatibility continues to be a challenge for all our air vehicles," Rudder said.[2]

What has happened since that time is the USMC is buying into the Reaper program and relying on a land-based remotely piloted vehicles to provide the support Marines would require for them at sea and from the sea operations. The Marines leased two Reapers from General Atomics since 2018 but then moved from leasing to buying the aircraft in 2021.

When Laird visited MAWTS-1 in November 2023, he learned how the force was practically moving ahead. MAWTS-1 is a place focused on training an integrated USMC force, not pursing systems that are simply "fairy dust" as one Marine put it to me. It is about how to make the force ready to fight tonight and to do so more effectively.

Laird discussed the integration of the Reaper into USMC operations with the LtCol Edgardo Cardona, the Executive Officer at MAWTS-1, who is a former DASC officer and current MQ-9A Reaper pilot.[3] LtCol Edgardo Cardona is one of the pilots where the Marines have created a new MOS, which is the 7318 MOS.[4] The Marine Corps Reaper unlike its Predator brethren is not armed so there is not a competition between remotely piloted or manned systems in terms of being trigger pullers.

U.S. Marines, assigned to Marine Corps Base Camp Pendleton, and U.S. Air Force Airmen, assigned to the 432nd Wing/432nd Air Expeditionary Wing, pose in front of an MQ-9 Reaper at Marine Corps Base Camp Pendleton, California, August 23, 2023. The agile combat employment (ACE) exercise known as Agile Hunter saw an Air Force MQ-9 remotely piloted aircraft land on Camp Pendleton, a Marine Corps base, for the first time ever. U.S. Air Force photo by Senior Airman Ariel O'Shea.

It is about enhancing the relevant ISR to provide for more effective insertion of force and enabling that force in terms of their operations. This is notably one the most significant changes since Laird last came to MAWTS-1 in 2020.

The career of the XO has paralleled that of the evolution of USMC experience in UAVs so that he is both a core officer in the evolution of USMC capabilities but has also embodied the transition from the Middle East and the Marines use of Shadow, Scan Eagle, Blackjack and K-MAX. He has been on the ground floor for the introduction of the Reaper to the Marine Corps.

The XO pointed out that his earlier experience at MAWTS-1 with UAVs, the focus was on deconfliction of the UAVs designed to provide ISR for the ground combat element. Now the focus is upon integration with the air element for the overall integrated operations.

The Reaper is working with the combat air elements in sharing a common operational picture and to enable those aircraft to have a view of the objective area prior to reaching it and to in turn to be able to enhance their ability to support the overall Marine Corps force being inserted into that objective area.

LtCol Cardona underscored that MAWTS-1 was working closely with the USAF on Reaper operations and sharing experience and understanding their different operational requirements as well.

He underscored: *Our goal with Reaper operations is to create a common operational picture enabling ground force commanders or maritime component commanders to make real time decisions based on a plethora of information that we're providing. And we're also focused on fusing different data links that are coming down from different services together to create that operational picture.*

He went to note that: *We see the MQ-9 as a good F-150 or a good reliable truck that can operate at long range and is reliable. But it is the payloads that are crucial to us and are ability to take the data generated by the payloads and use our digitally interoperable systems to distribute the data throughout the MAGTF.*

When he came to MAWTS-1 in 2020, he underscored:

We needed to figure out how to shape an MQ-9 program within the WTI focus of MAWTS-1. Training is a key piece in standing up a new capability and at MAWTS-1, it is about integrated MAGTF capability. We are not training a stand-alone force.

The XO noted that they reached out to the Air Force to help validate their initial MQ-9 training approach, and now they share lessons learned and share training slots when appropriate.

LtCol Cardona underscored: *We are working with the ACC and Headquarters USMC to set up an exchange program to foster the expertise required.*

Any time I have excess capacity, I will take an Air Force student and make them a WTI. And then they will return to the Air Force community. We have a Marine currently in the Air Force 26 Weapons School training program who will graduate in December.

So now we have a WTIs in the Air Force, and we're going to have USAF weapons school graduates in the Marine Corps who are Marines. And it fosters a lot of TTP development, a lot of great relationships with the Air Force.

He noted as well that they are tied in with the operational test community via VMX-1. They want to do integrated testing on new sensor suites and to be able to provide user input prior to the decision of what exactly gets produced and acquired.

The Marines will need to become major players in autonomous systems – airborne, and surface and below surface systems—the Reaper is beginning the process. But certainly, the unique integrated mission sets the Marines work through a MAGTF will drive innovation which the joint force needs to note.

Col Purcell's Perspective on the Impact of the Coming of the CH-53K

November 22, 2023

During Laird's visit to MAWTS-1 during the first week of November 2023, he had a chance to talk with Colonel Eric Purcell, the CO of MAWTS-1 about the coming of the CH-53K to the USMC.

This is the third new air system Laird has seen coming to the USMC since he has been coming to Yuma, but because it doesn't look as different as the other two did from their legacy ancestors, it is often not fully realized how important it will be for the USMC and the joint force.

Purcell is the first CH-53 pilot to be the CO of MAWTS-1 which is propitious as the CH-53K has been part of this year's WTIs at MAWTS-1. He has more than 3000 hours on the CH-53E and 130 hours on the CH-53D. He has had two deployments to Afghanistan and two to Iraq, and additional visits to both countries as well.

He noted that he wished they had not called it the CH-53 for the CH-53K is so different from the legacy aircraft. It is designed to fit into the deck space of an CH-53E and to have a reduced footprint for its maintenance as well.

But the big difference is associated with the broader changes across the Marine Corps. When Laird was last at MAWTS-1 in 2020, they were starting to work on how to enhance the deployability and mobility of the Marine Corps and to do so in formations smaller than the traditional MAGTF.

During this visit, Laird's discussions with the department heads underscored how much work they have done in terms of doing expeditionary basing, innovations in Forward Refueling and Re-Arming points and ways to reduce the signature of the deployed force.

The CH-53K, in Col Purcell's view, contributed to that in a significant way. He focused on the ability of the King Stallion with its triple hooks to carry significant loads to operating locations without having to land and be on the ground for the time necessary to unload from the interior of the aircraft.

Col Purcell pointed out that the aircraft could carry significant fuel loads – 54,000 pounds of fuel -- to locations the F-35B might operate from and could do so with external lift rather than having to land.

Both the Osprey and the heavy lift helo could carry fuel inside and work as fuel providers to aircraft at a FARP. But being on the ground for significant time to do this exposed the aircraft to much greater risk than coming in and dropping off fuel from their external three hook system.

He pointed out that the legacy aircraft two hook system could lead on occasion to "uncommanded" load releases whereby the system on the aircraft would not be able to judge correctly whether loads on the hooks were compromising the safety of the aircraft. Systems on the aircraft prioritized aircraft safety over carrying loads and might jettison a load.

The CH-53K's systems can correctly determine whether the load being carried by the aircraft affect the center of gravity of the aircraft, which is central to its security, and can make more accurate decisions with regard to the safety of the aircraft.

MAWTS-1

U.S. Marine Corps Col. Eric. D. Purcell, Marine Aviation Weapons and Tactics Squadron One (MAWTS-1) commanding officer, conducts an operations brief in support of a salvage and recovery exercise during Weapons and Tactics Instructor (WTI) course 2-23, at Auxiliary Airfield II, near Yuma, Arizona, March 17, 2023. U.S. Marine Corps photo by Lance Cpl. Ruben Padilla.

He noted that the load carrying capacity of the aircraft meant that it could carry an Osprey which might be in a location where it could not get repairs needed to fly safely to a location where it could be repaired. Some of the weight, such as the seats, would have to be removed to do so, but it could be done.

Col Purcell underscored that in Afghanistan and Iraq many of the missions which the CH-53E did were medium lift. The CH-53K is optimized for heavy lift and both the Marine Corps and the joint force need to focus on its unique capabilities to support distributed logistics as no other rotorcraft can do in the force. It is optimized for heavy lift, and it is important to capitalize on its unique capabilities.

The CH-53K can be part of a logistic chain involving cargo aircraft like the C-17, the C-5 and the C-130, in that it can carry 463L pallets and work with fixed wing cargo aircraft to transfer their pallets to the Super Stallion and then deliver them in places only a rotorcraft can go.

The new motors on the King Stallion allow it to operate in conditions where one would not want to operate an aging CH-53E fleet. The power margins of the new aircraft are much greater than the legacy aircraft.

Col Purcell concluded: *The force will see the impact of the revolutionary design of the CH-53K to carry heavy loads long range and to enhance significantly the logistical capability of the force and to move in and out of objective areas more rapidly than the legacy system.*

Operating from the Seafloor to the Heavens: MAWTS-1 Works the Spectrum Warfare Challenge

November 27, 2023

During Laird's visit to MAWTS-1 in 2018, the shift from operating in the land wars to dealing with challenges which could be posed by peer competitors was clearly underway. At the center of the shift was clear recognition of the need to deal with the ongoing challenge of operating in contested electro-magnetic environments.

In his meetings with PACFLEET this past April, a key aspect of the challenge facing the force is clearly spectrum warfare or signature management and deception.

His takeaway from discussions at PACFLEET was very clear. Success for distributed maritime operations requires not only assured command and control but the tissue of ISR systems enabling distributed fleet operations and adding the key element of deception through various counter-ISR systems as well.

In effect, fleet distribution built on a kill web effects infrastructure is being combined with what might be called a wake-a-mole operational capability. 'You can't target me, if you can't find me.'

As one key Navy leader put it to Laird: "Counter-ISR is the number one priority for me, to deny the adversary with to high confidence in his targeting capabilities. I need to deceive them and to make a needle look like a needle in a haystack of needles. It is important to have the capability to look like a black hole in the middle of nothing."

MAWTS-1 with its work on FARPs, force distribution, new ways to do C2 such as from systems operating out of the back of Ospreys, signature management and deception is clearly working along the same lines as PACFLEET.

Laird had a chance to discuss this approach with Major John Edwards, the Spectrum Warfare Department Head, when visiting MAWTS-1 in November 2023. Edwards background is from operating in the last remaining specialized USMC airborne electronic warfare platform, the Prowler.

Edwards highlighted their work in preparing the force for the spectrum management challenges they experience and will experience within a contested electronic warfare environment. Their focus is upon ramping up the "red" threat against "blue" to prepare the force for the experience they have when operating in a contested environment.

They focus on showing the operational elements of the force the kind of spectrum signature they are creating when they operate to better understand the nature of spectrum warfare and how that affects their lethality and survivability in operations.

And by knowing that one can create ways to think about signature deception as well as pointed out by PACFLEET.

The focus now is upon having the entire operational force "electronic warfare" ready, which requires consideration as well for bringing the kind of threats which red forces can be brought by an adversary to contested operations. This is always going to be a work in progress as both "blue" and "red" work the dynamically changing electronic magnetic spectrum embodied in the operating forces.

We discussed an issue as well which Laird has observed in both allied and U.S militaries, namely the paring down of specialized aircraft to perform the EW mission and the question of whether there is enough capability remaining for the specialized domain which EW or tron warfare really constitutes.

U.S. Marine Corps aircraft assigned to Marine Operational Test and Evaluation Squadron One takeoff from a landing zone during a Spectrum Warfare Department (SWD) training event near Yuma, Arizona, Aug. 23, 2023. SWD Marines conducted a training to intercept, identify, and locate or localize sources of intentional and unintentional radiated electromagnetic energy for the purpose of immediate threat recognition, targeting, planning and conduct of future operations. U.S. Marine Corps photo by Lance Cpl. Ruben Padilla.

Major Edwards thought this was a valid concern and observed that the Navy currently through its Growlers provides a joint capability. He noted the USAF has officers involved with Growlers. Growlers come to MAWTS-1 as well to work with the range of capabilities which the USMC deploys.

But Edwards felt that there needs to be a joint force commitment to expertise in this area beyond simply relying on the Growler community.

Part of the problem is simply the need to have a robust capability in this area which can inform the force with regard to training standards necessary for effective operations.

If it was felt correctly that spectrum management is now a thread running through offensive and defensive operations, where will the expertise be to inform the operational and training requirements?

"Our excellence is a key part of deterrence"

November 27, 2023

During Laird's November 2023 visit to MAWTS-1, he had a chance to talk with Major Kyle "Elton" McHugh, tactical air department head at MAWTS-1.

McHugh was originally a Harrier pilot who transitioned to the F-35B and served for three years with the first forward deployed F-35B squadron, VMFA-121.

The F-35 is a key part to answering the question behind how the USMC integrates with the joint force and supports the maritime fight.

The Marine Corps F-35Bs in Japan provide fifth-generation aircraft to the Marine Corps and Joint Force in the first island chain, operating off a variety of basing options, and work closely with the USAF, USN, and allied Pacific Air Forces.

The aircraft shares common sensors, decision making systems, weapons and so on with a common U.S and allied combat fleet. In discussing the way ahead for USMC integration with the joint force, both the F-35 and the Osprey are often overlooked as key stakeholders in both current capabilities and the future force.

Laird asked McHugh about his time in Japan and his perspective on working with the Japanese. He commented: "I think our enemies fear our excellence and that is a key part of deterrence.

"Working with the Japanese as they stood up their own F-35s, we shared a common mission and a common passion to defend our nations and our way of life. Our adversaries cannot ignore that commitment and the quality we bring to the fight."

Throughout our discussion, Major McHugh emphasized how the Marines have worked with the joint and coalition force on integrability of the F-35, enabling a more lethal and capable F-35 enterprise.

U.S. Marine Corps Maj. Kyle McHugh, a pilot assigned to Marine Fighter Attack Squadron 121, currently attached to Marine Medium Tilt-Rotor Squadron 265 Reinforced, 31st Marine Expeditionary Unit (MEU), departs the flight deck of amphibious assault ship USS America (LHA 6), in the East China Sea, June 24, 2021. Photo by U.S. Marine Corps Staff Sgt. John Tetrault.

He noted that: *There has been a concerted effort by MAWTS-1 to work with the Navy and Air Force weapon schools on F-35 integration. All the weapon school instructors meet in person twice a year. The goal is to standardize and shape a tri-service TTP manual, including Australian and British partners as well.*

We also discussed the coming of the Reaper and he emphasized how its inclusion is helping to learn about and shape operations in the challenging maritime environment.

He noted that many of the students who come to WTI do not have deep knowledge of the maritime environment and Reaper data is helping in that learning process.

He noted the last Harrier class came to this year's WTI and the Hornets are soon to follow. The all F-35 Marine Corps TACAIR element will be a key part of the joint and allied integration efforts as evidenced by the work on a common training manual by the weapon school instructors.

Major McHugh concluded: *At MAWTS-1 we are focused on tactical excellence. That level of competence is critical to deterrence. The events we do at MAWTS-1, both live and simulated, are executed with distributed joint partners, empowering mission commanders to compete against the evolving threat.*

ISR, Mission Planning and Enabling a Distributed Force: An Ongoing Challenge

November 28, 2023

Laird spent parts of March and April this year in Australia and from, there flew to Hawaii where he visited PACFLEET and PACAF.

The Australians and the American commands are both working to build a path to enhance deterrence of China and then augment capabilities which work to reinforce that path. Force distribution, greater allied interoperability, significant C2 and ISR capabilities enabling the distributed force to operate as a kill web with enhanced capabilities to confuse adversary targeting are the key elements for reshaping the current force.

Based on this effort, acquisition choices will either help or hinder augmentation of core capabilities to further realize this approach.

A key challenge in mission success is working effective ways for higher level commanders to work effectively with forces at the tactical edge which increasingly have ISR capabilities better than the commanders have with regard to that particular operational area.

How to proceed? As one Navy officer put it:

Higher headquarters must be able to see, understand, monitor, and adjust tactical headquarters that own battlespace and missions throughout the theater.

Higher headquarters must have the ability to see, understand and occasionally direct. But those headquarters must have borders so the tactical commanders can exercise their own creativity to deliver the fires and effects where they are operating.

The higher headquarters may have access to better information and when it does it needs to have the ability to reach out to the tactical level to tell them to do or not do something associated with the larger political and strategic picture.

He felt that they were making significant progress in

commanding a distributed force, which is a core element of shaping a force capable of deterrence in the Pacific.

We are capable of commanding from various locations and can be able to see and understand how to command in the battlespace dozens of ships, hundreds of aircraft, thousands of personnel. We are capable of seeing, understanding, and deciding what is going on in the battlespace, and tracking the enemy force using exquisite means way beyond a grease pencil and a radio call. We can and do so through links and sensor from the sea floor to the heavens.

F-35B Lightning II with Marine Fighter Attack Squadron VMFA-211 and the United Kingdom's 617 Squadron, both Carrier Strike Group 21, are secured to the flight deck aboard HMS Queen Elizabeth at sea on May 06, 2021. Credit: UK Ministry of Defence

During Laird's visit to MAWTS-1 in November 2023, he had a chance to talk about how the intelligence aspect of working in mission command and the evolving ISR environment was being experienced by a USMC intelligence officer. He met with Capt Liggett, a MAGTF Intelligence Officer and a U.S. Naval Academy graduate whose generation will experience very significant dynamic decades of change of both ISR and counter-ISR dynamics.

Prior to coming to MAWTS-1, her experience has been in supporting fixed wing aircraft and their mission planning. She has worked with VMFA-211 and served on the Queen Elizabeth during the Marines operating their F-35Bs from that ship.

This is a unique experience for the British carrier was built especially to operate the F-35s and has information warfare and intelligence facilities built on the ship especially for F-35 capabilities. When Laird visited the ship when it was being built in Scotland, he saw those facilities being built and their functionality was explained to him at the time.

Capt Liggett onboard HMS Queen Elizabeth.

Capt Liggett indicated that coming to MAWTS-1 has broadened her experience and allowed her to work the intelligence function for the mission planning of the entire air capability of the USMC, including rotorcraft and the new addition to the force, the Reaper.

Laird highlighted that with the kind of ISR at the tactical edge which forces have this changes the dynamic between intelligence that comes from other sources and the dialogue that goes on between operators and the intelligence function.

Capt Liggett discussed that dynamic in the following terms:

We need to communicate back and forth to ensure that we have most updated information to pass on. We are active participants in this process and working to ensure that from a collection point of view we are looking in the right locations and refining information with regard to the operational area.

The challenge is to pass data effectively to the right people at the right time and make sure we have access to resources that enables us to do that. We are also informing our pilots of what we can know and what we can't from the particular resources available to us at a particular time.

She indicated that: *We are generally cueing the intelligence data and the operators at the tactical edge are then taking that with what they see in real time to prosecute targets.*

With the growing capability of ISR inherent in the force at the tactical edge – with F-35s and other local intelligence capabilities – the dialogue with the intelligence analysts and sources beyond the tactical edge with those at the tactical edge is a key part of shaping operations going forward.

The USMC Ground Combat Element and Shaping a Way Forward: Challenges to be Met

November 30, 2023

During Laird's visit to MAWTS-1 in November 2023, he had a chance to talk with Maj Scott Mahaffey, the Department Head for the Ground Combat Department.

Maj Mahaffey has been at MAWTS for three years but prior to that has had several deployments, including Afghanistan, with an SP-MAGTF, which involved operations in Africa. He also participated in a number of exercises such as Trident Juncture in Norway in 2018, the first exercise with the Indians, in the Tiger Triumph Exercise, and multiple exercises in the Philippines.

This background provided a very good preparation for taking on the demanding task of working the training of the GCE as it works its transition from more traditional ground operations associated with the land wars to a wider range of force insertion missions.

According to Mahaffey: *Predominantly our portion of WTI we call our Air Assault and Fires Integration course. It is a heavy flavor from a ground and ACE perspective of how to execute air assaults from battalion sized raid to a platoon-sized lift.*

The fires integration piece at least from our student's perspective is focused on fires integration with air-delivered weapons and how they can mesh and integrate with ground fires not necessarily in the close in battlespace but in the deep enclosed battlespace.

U.S. Marine Corps Capt. Scott Mahaffey plans the day's movements with Indian Army Soldiers during exercise Tiger TRIUMPH in Kakinada, India, on Nov. 19, 2019. Tiger TRIUMPH improves U.S.-Indian partnership, readiness and interoperability. It gives the U.S. Marine Corps and Indian forces the opportunity to work together, exchange knowledge and learn from each other on a range of military operations such as humanitarian assistance disaster relief and amphibious operations. Mahaffey is the commanding officer of Easy Company, 2nd Battalion, 2nd Marine Regiment, on a Unit Deployment Program to 4th Marine Regiment, 3rd Marine Division and is a native of Orlando, Fl. U.S. Marine Corps photo by Cpl. Jacob Hancock.

Our students are typically captains at the company level, some senior enlisted, gunnies or master sergeants, with a sprinkling of lieutenants. They have focused on how to accomplish a company objective with close in fires, and we focus on training to give them knowledge of fires support at higher echelon support.

We are focused on training which enhances their knowledge of the broader battlespace and how F-35 or HIMARS fires provide support for a broader engagement in the battlespace.

When Laird was previously at MAWTS in 2020, there was

significant concern with deploying Marines to remote locations, even if new longer-range weapons will come on line in the next few years, with the question of fires authorities. In the MAGTF construct, the question of who gives fires authority is clear—in the new direction to have smaller groups of Marines deployed throughout the battlespace, that question is not yet resolved.

The other issue is that GCE is training in the absence of the longer-range weapons which are not yet in place. They are using data from weapons development efforts, a step up from briefing slides, to work with notional weapons, NEMESIS being a case in point.

The challenge facing the GCE is seen in this USMC 2021 description of **NMESIS**.

The Navy/Marine Corps Expeditionary Ship Interdiction System successfully hit its target in support of Marine Corps Forces, Pacific, during Large Scale Exercise 21 Aug. 15, 2021. The exercise showcased the U.S. maritime forces' ability to deliver lethal, integrated all-domain naval power.

LSE 21 was a live, virtual and constructive scenario-driven, globally-integrated exercise with activities spanning 17 time zones. LSE 21 applied and assessed developmental warfighting concepts that will define how the future Navy and Marine Corps compete, respond to crises, fight and win in conflict.

The Marine Corps' NMESIS will provide the Marine Littoral Regiment with ground based anti-ship capability to facilitate sea denial and control while persisting within the enemy's weapons engagement-zone, and LSE 21 provided a venue for the program team to validate some of those concepts.

"This scenario is representative of the real-world challenges and missions the Navy and Marine Corps will be facing together in the future," said Brig. Gen. A.J. Pasagian, commander of Marine Corps Systems Command. "This exercise also provided an opportunity for us to work alongside our service partners to refine Force Design 2030 modernization concepts."

SINKEX, the exercise scenario involving NMESIS, provided a testing environment for new and developing technologies to connect, locate, identify, target and destroy adversary threats in all domains, culminating in the live-fire demonstration of the naval strike missile against a sea-based target.

During the exercise, forward-deployed forces on expeditionary advanced bases detected and, after joint command and control collaboration with other U.S.

forces, responded to a ship-based adversary. Simultaneous impacts from multiple, dispersed weapons systems and platforms across different U.S. services—including NMESIS—engaged the threat.

NMESIS integrates established, proven sub-systems, such as the Joint Lightweight Tactical Vehicle Chassis, the Naval Strike Missile and the Fire Control System used by the Navy for NSM.

"From an acquisition perspective, NMESIS started a little over two years ago," said Joe McPherson, long range fires program manager at MCSC. "We've been able to rapidly move [on developing and fielding this system] because we're leveraging existing NSM and JLTV subsystems."

Because NMESIS is not yet a fielded capability, engineers from MCSC managed the fire control piece of the system during the exercise. Marines, however, were able to practice maneuvering the system and validating the system's interoperability with their Naval and Air Force partners.[5]

The basic focus of the Ground Combat Department at MAWTS-1 is upon company training and awareness of higher echelon support, both actual and projected such as the case with NMESIS.

Maj Mahaffey indicated that simulators are a key part of the training regime, which allows for lessons learned to be taken away and reflected in the simulators which are located at the Marine Corps bases as well. This allows for working the standardization aspect of the training development process.

But they don't have a simulator for a new system such as NEMESIS and are working with a notional projected system as part of their training regime. They are training in terms of working the timelines to execute a strike with the notional system to support an external fires authority.

The reality for the GCE is that currently they have organic fires which allow them to operate in the close in fight. Longer range fires are in development, with HIMARS at an outer range of 300Kms being the exception.

But by and large the GCE projected in remote locations is largely capable of providing defensive capability, protecting logistical locations, or radars or items of interest, not being part of an offensive punch for the joint force.

It is a work in progress.

C3 and the Way Ahead for the USMC

December 4, 2023

C3 is the key tissue allowing for the shaping of a distributed force which can be integrated to create the desired combat effect. During his November 2023 visit to MAWTS-1, Laird discussed the way ahead in this crucial area with Major Christopher Werner, the C3 Department Head.

Werner is a DASC marine, as is the current XO of MAWTS-1.

These are Marines who focus on providing the direction for air operations supporting the ground forces, and hence are key players in shaping an integrated ground maneuver force.

Major Werner began by describing their key role, notably working with the USAF, in providing fire support for the ground forces in Afghanistan. He noted that they were able to operate within a MAGTF construct to provide significant support for the ground forces in counter-insurgency operations.

But with the shift to preparing for combat operations against peer competitors, the focus has shifted both for the air element and the focus of C3 Marines on force integration. The air element is now returning to air-to-air combat as well as air defense as key missions, both of which were not the focus of attention in the counter-insurgency wars.

And now the C3 effort for the USMC needs to shift from a primary focus on integration within the MAGTF to working with the joint force and using C3 to integrate relevant joint force elements to create the desired effect.

Werner noted that their cooperation with the USAF evident in Afghanistan was going forward, but there was a renewed emphasis on working with the Navy on new ways to do force integration, to which C3 needed to provide the integrating tissue.

But this was a work in progress.

Major Werner noted that officers of the same rank of his in the Navy are pushing for more effective C3 between the services, but major problems remain in terms of working with the C3 systems of the carrier task forces.

He underscored: *That is something we need to work out if we can operate as an inside force with the carrier strike groups.*

C3 for the joint force as seen with the Navy MISR officers is essentially sensor-shooter integration over a kill web.

As Major Werner put it: *What we teach in our courses is the importance of being able to take data from whatever joint or coalition forces sensors are relevant to us and blending them into data enabling shooters and fires control decision makers.*

I think that makes our community such an interesting one to work within today, notably as the joint services pursue ways to do joint command and control to create desired combat effects at the tactical edge.

An important focus which would enable the Marines is building on the ARG-MEU and its amphibious ships. In our book on the maritime kill web, we underscored how we envisaged a very dynamic future for such a force.

U.S. Marines with Command, Control and Communication, Marine Aviation Weapons and Tactics Squadron One (MAWTS-1), board an MV-22B Osprey aircraft during an offensive air support exercise, part of Weapons and Tactics Instructor (WTI) 2-23, at Marine Corps Air Station Yuma, Arizona, April 4, 2023. U.S. Marine Corps photo by Lance Cpl. Ruben Padilla.

As we argued: *There is no area where better value could be leveraged than making dramatically better use of the amphibious fleet for extended battlespace operations. This requires a re-imaging of what that fleet can deliver to sea control and sea denial as well as Sea Lines of Communication (SLOC) offense and defense.*

Fortunately for the sea services, such a re-imaging and reinvention is clearly possible, and future acquisitions which drive new connectors, new support elements, and enhanced connectivity could drive significant change in the value and utility of the amphibious fleet as well.

In addition, as the fleet is modernized new platform designs can be added to the force as well. And as we will address later in the book, this entails shaping variant payloads as well to be delivered from a distributed integrated amphibious fleet.

As building out the evolving fleet, larger capital ships will be supplemented and completed with a variety of smaller hull forms, both manned and autonomous, but the logistics side of enabling the fleet will grow in importance and enhance the challenges for a sustainable distributed fleet.

That is certainly why the larger capital ships – enabled by directed energy weapons as well – will see an enhanced role as mother ships to a larger lego-like cluster of smaller hull forms as well.[6]

And as maritime autonomous systems come on line, amphibious ships are well positioned for mother ship functionality in terms of launching and leveraging air and sea autonomous systems.

Crafting a Sustainable Distributed Force: Maintenance and Logistics Challenges

December 5, 2023

Each of the services is seeking ways to distribute their force for survivability and presence, but at the same time working through a joint lens to enhance lethality.

But the elephant in the room is sustainability, maintainability, and broader considerations of logistics support.

MAWTS-1

U.S. Marines with Marine Heavy Helicopter Squadron 361 (HMH-361), Marine Aircraft Group 16, 3rd Marine Aircraft Wing, perform maintenance on U.S. Marine Corps CH-53E helicopters assigned to Marine Aviation Weapons and Tactics Instructor (WTI) course 1-24 at Marine Corps Air Station Yuma, Arizona, Sept. 20, 2023..Marine Corps photo by Lance Cpl. Ruben Padilla.

When visiting PACFLEET and PACAF this past April, one of the key subjects we discussed was the problem of how to have a sustainable distributed force.

- How to do you support a distributed maritime force?
- How do you support an Air Force doing agile combat employment?
- How do you maintain a force which is distributed?

It is no shock that when the Marines are focusing then on EABO and other approaches to force distribution that they face similar problems. One can demonstrate a FARP in which the F-35 receives fuel, but what if it needs the right kind of maintainer and the parts that are needed when it wishes to leave the FARP?

- How do you move ammunition and weapons to keep a sustained moving EABO force?

- Since this is not done by pixie dust, what support systems are available?
- How are the personnel trained and available to do such operations?
- Since the trend is away from iron mountains of weapons and material how do you create a chessboard of support for the force at the tactical edge?

To do so will require significant organizational change in DoD and the shaping of a truly joint sustainment system.

U.S. Marine ordinance technicians assigned to Marine Aviation Weapons and Tactics Squadron One, prepare to load advanced precision kill weapon systems during a forward arming and refueling point exercise, part of Weapons and Tactics Instructor (WTI) course 1-24 at Landing Zone Bull Attack, near Chocolate Mountains, California, Oct. 13, 2023. U.S. Marine Corps photo by Cpl. Alejandro Fernandez.

But the services are moving out on ways to distribute force in advance of any such changes.

This is obvious in a place like MAWTS-1 where the focus is upon the standardization of training for operations.

How to standardize maintenance and sustainment for the EABO focused Marines?

To be blunt this is a struggle and an effort in progress, but the resolution of this challenge is at the joint sustainment level not simply at the USMC or U.S. Navy level. And joint solutions can be crafted using the air systems and autonomous systems coming in the near to mid-term including USVs and UAVs.

Laird discussed the current situation with regard to standardization of maintenance and weapons support with three MAWTS-1 Marines. Major James J. Lay is the head of aircraft maintenance at MAWTS-1. Captain Jonathan R. Caruthers oversees aviation ordinance support to the aircraft operating at WTI. And Captain Tyler Thomsen who is the director of the advanced aircraft maintenance course at MAWTS-1.

Laird had a number of take-aways from my discussion with these three very capable officers. Laird is not holding them responsible for how he interpreted the discussion but am crediting them with insights with regard to the whole transition challenge.

- The first is simply how daunting their task really is. MAWTS-1 is not an owner of aircraft. It is a facilitator of integrated operations which means that aircraft come from the various USMC air wings, fly into Yuma and within one week, they have to make the aircraft ACE ready. They then must organize the maintainers as part of a command-wide training effort to shape standardization across of the force. Maintainers come from the Marine Air Wings. It is in MAWTS that best practices are brought together, allowing the standardization of those practices throughout the USMC. This obviously is ongoing learning process which is driven by maintenance experience of the air wings in operations which then gets standardized at MAWTS-1.

- Second, the template for provision of weapons has been the iron mountain at a base or support facility then moved out to the areas of interest. But this template needs to be adjusted in several ways. There needs to be a ramp up of weapons supply. There need to be new methods and procedures to distribute weapons across a distributed force. The work which MAWTS-1 has been doing regarding FARPS/EABOs provide a demonstration of how challenging movement of weapons over a distributed force operational area is and the question of how to work this remains to be resolved. The problem in part rests in acquisition. The Marines acquire ground-based ammunition: The Navy acquires weapons for the air element. How to ensure a common adequate flow of weapons to a USMC expected to support the Navy in new ways?

- Third, there is the need clearly to provide a different generation of weapons for the use of a distributed force designed to deliver integrated fires at range and distance required. But what will be the new template to distribute such forces to an agile combat force?

In short, MAWTS-1 is an incubator for testing the real world for new approaches and certainly along with the Navy and the Air Force's weapon schools will attack this challenge in real world terms.

Expanding the Assault Support Mission to a Broader Mission Set

December 8, 2023

During Laird's November 2023 visit to MAWTS-1, he had a chance to meet with the Assault Support Department Head, Maj Nicholas Peters, to discuss the activities of the assault mission training part of MAWTS-1. While the traditional image of assault support remains a key one – transporting Marines to a place of embarkation and ready to fight – there is growing emphasis on longer range missions and on broadening the mission set.

The Osprey and KC-130 in SP-MAGTF operated at distance. This experience is being folded into training for long range missions in the Pacific. The other members of the assault force – the rotorcraft including heavy lift – are not built for the range and speed of a combined KC-130J-Osprey mission set.

But working the broader assault package in areas of interest remains a key bread and butter capability of the USMC, and continued press of events such as the Middle East certainly reminds one of the necessities of ensuring that the Marines are range for a spectrum of operations, not just somebody's pet rock.

Earlier this year, Laird interviewed Col Marvel who identified a range of adaptations which the Osprey is currently going through to support the joint force.

Col Marvel underscored that expanding the mission set for the Navy's CMV-22B was certainly possible but was not in his domain of responsibility.

But the USMC is clearly expanding the payloads carried by the MV-22B which supports distributed operations, and if the three services which operate the aircraft found ways to expand their ability to cross-service each other's aircraft, they would be able to enhance such operations.

As Col Marvel put it: *The Osprey provides unique speed and range combinations with an aircraft which can land vertically. It is a very flexible aircraft which could be described as a 'mission-kitable' aircraft.*

The Osprey has big hollow space in the rear of the aircraft that can hold a variety of mission kits dependent on the mission which you want the aircraft to support.

He emphasized that with a variety of roll-on roll-off capabilities with different payloads.

We can add the specialists in the use of a particular payload along with the payload itself to operate that payload, whether kinetic or non-kinetic, whether it is a passive or active sensor payload. We need to stop thinking about having to put the command of such payloads under the glass in the cockpit and control those payloads with a tablet.[7]

Maj Peters indicated that at MAWTS-1 they have expanded the mission sets for their Ospreys to embrace C2 and ASW efforts.

With regard to the C2, roll on roll of capability can provide for a variety of joint force enablement and support missions. With regard to ASW, the Osprey can deploy sonobuoys in support of the Navy's

ASW mission as well, and they have exercised such capability at MAWTS-1.

A key aspect of the new emphasis for Osprey training being performed at MAWTS-1 is the TRAP mission for the Navy. Obviously, there have events in the past such as the pilot rescue in Libya which highlighted how the speed and range of the Osprey provides unique TRAP mission capabilities.

But now with the focus on Indo-PACOM and the concern for loss of aircraft in a contested operation, it is important for the U.S. Navy to rely on the speed and range of the Osprey to support the TRAP mission.

Here the Osprey community is working hoisting methods to provide for the mission, and this has become part of the training conducted by the Assault Support Department.

Landing on USS America was during Maj Peters time with VMM-265 and from April 2020. Credit: USMC

The Osprey is flown by the USAF and Navy as well, which leads to a kind of built in joint integration in terms of a common operator pool across the services.

Peters indicated that a USAF Osprey pilot was an instructor in his department and taught the students how the USAF using its Ospreys and operated them differently from the USMC.

And Laird's visits to North Island with the CMV-22B squadrons certainly underscores that with the Navy operating their version of the Osprey, there are significant opportunities for working maritime integration at a very fundamental support and assault level as well.

Opening the Aperture on Force Integration

December 11, 2023

The strategic shift from the priority on the land wars to dealing with a world of multi-polar authoritarians and a diversity of contingency operations requires significant operational change in the force.

Because MAWTS-1 is focused on standardized training for the operational force, they have an open aperture to be able to incorporate operational change but not from briefing slides and wargames, with the force that the Marines that is ready to fight tonight.

As the end of course video for WTI-1-24 opens with: "It is not a question of if the Marine Corps will go into combat, it is only a question of when."

With this perspective the future is now but done with a perspective to be able to add real capabilities wherever the Marines can find them.

With this perspective the future is now. As one former CO of MAWTS-1 told us: "The Marines are the ultimate scavengers."

A graphic illustration for Tactical Air Control Party, Marine Aviation Weapons and Tactics Squadron One who assisted in close air support exercises during Weapons and Tactics Instructor (WTI) course 1-24, at Marine Corps Air Station Yuma, Arizona, Oct. 25, 2023. U.S. Marine Corps graphic illustration by Lance Cpl. Emily Hazelbaker.

In November 2023, Laird had a discussion with Major Green who is a UH-1Y pilot Joint Terminal Attack Controller Evaluator (JTAC-E) as well as the Department Head of the Tactical Air Control Party (TACP) Department. He explained that when he first came to MAWTS, the department was called the Air Officer Department.

But this name did not reflect the reality of the personnel trained by the department. Many of the personnel involved are pilots and after training at MAWTS-1, and they go back to their units as Forward Air Controllers.

At the completion of their time as a Forward Air Controller or Air Officer, they return to their squadron. The other half are enlisted Marines who serve as JTACs or Joint Terminal Attack Controllers.

As we talked, it was clear that the experience of the land wars whereby JTACs controlled Close Air Support missions was changing as the USMC focused on maritime operations and deep interdiction missions. How to do this shift successfully is a work in progress, but the role of the ground controllers is changing. Historically, they have not had access to Link-16 which is a key system for air integration and situational awareness in the Joint Force.

The C3 department has access to a variety of means to integrate data, but the challenge now is to push an ability to do similar integration to the ground controllers.

Or put in other words, legacy CAS was the focus of the ground air controller: their role now was to work as an integrative element between the GCE being distributed in the battlespace with the air element which can provide sensing, strike, and communication links for the distributed GCE.

Major Green indicated that there was an increasing focus on new simulation capability for training at MAWTS-1. The legacy systems are too limited to meet the demands for training the force for distributed and deep strike operations.

With new prototype simulation capabilities in their hands, Major Green saw the future development of LVC as an important tool in shaping a way ahead to train to new concepts of operations as well.

He noted that during his time at MAWTS-1 (nearly three years), there has been enhanced focus of attention on the maritime domain and ways the USMC operates in that domain. He noted that there was "more of a maritime flavor to our training efforts."

In short, the USMC is trying to figure out what FARPs/EABOs actually mean in the evolving combat environment, and this obviously affects the role of the TACP elements in linking together the air and the ground elements going forward.

ROBBIN LAIRD & EDWARD TIMPERLAKE

The Challenges of Working Expeditionary Advanced Base Operations

December 8, 2023

During Laird's last visit to MAWTS-1, the training effort was clearly focused on ways to enhance force mobility and lethality. There is a clear challenge in trying to determine how to position a distributed force, how to size it in order to have meaningful force capability along with enhanced survivability.

- How do you position the force?
- How do you organize the force?
- How do deploy and move the force?
- How do you find ways to reduce signature management of such a force?

In other words, at MAWTS-1, they are not wargaming Expeditionary Advanced Base Operations, they are working to determine how most effectively to do so and yet have meaningful combat effects. Not an easy task.

Not finished and a work in progress to determine movement, logistics support, C2 and fires solutions. It was clear when the MAGTF was the organizing principle but not so much with using EABOs as an organizational construct.

EABOs are described as follows: *Expeditionary Advanced Base Operations is a form of expeditionary warfare that involves the employment of mobile, low-signature, operationally relevant, and relatively easy to maintain and sustain naval expeditionary forces from a series of austere, temporary locations ashore or inshore within a contested or potentially contested maritime area in order to conduct sea denial, support sea control, or enable fleet sustainment.*

EABO support the projection of naval power by integrating with and supporting the larger naval campaign. Expeditionary operations imply austere conditions, forward deployment, and projection of power.

EABO are distinct from other expeditionary operations in that forces conducting them combine various forms of operations to persist within the reach of adversary lethal and nonlethal effects.

It is critical that the composition, distribution, and disposition of forces executing EABO limit the adversary's ability to target them, engage them with fires and other effects, and otherwise influence their activities.[8]

During Laird's visit in 2020 to MAWTS, he talked with Maj Steve Bancroft, Aviation Ground Support (AGS) Department Head, about their efforts working this problem set.

Bancroft focused on the various ways they were working enhanced force mobility, but a knotty problem was how to speed up the creation and withdrawal from EABOs.

Laird continued this discussion during his 2023 visit with the current AGS Department Head and USMC combat engineer, Maj Justin Atkins.

Atkins noted that in his deployments to date, they had not really focused on signature management. When fighting the land wars, signature management was not a key issue.

U.S. Air Force C-130 Hercules aircraft departs from a forward arming and refueling point during Assault Support Tactics 4 (AST-4), part of Weapons and Tactics Instructors (WTI) course 1-24, at Sandhill, Marine Corps Air Ground Combat Center, Twentynine Palms, California, Oct. 24, 2023. U.S. Marine Corps photo by Cpl. Alejandro Fernandez,

But when dealing with more advanced adversaries, obviously operations in the electro-magnetic spectrum had a key effect on the movement and operation of forces.

With regard to EABOs, the question of how to manage forces across the combat chessboard is clearly affected by signature management and the need to organize force in ways to reduce it or to mask it. He noted that most of AGS activities are focused on FARP operations as the means to do EABOs.

They have worked multiple configurations of FARPs to do so but have not found an optimal solution. *We are building small tactical teams and exploring ways to sense, communicate, and to operate in the battlespace with mobility. But how to ensure that such teams have the desired effects?*

He noted that they work with the spectrum warfare department to do two things.

- First, they work with them to reduce their spectrum signature footprint.
- Second, they are working as well to copy that footprint to provide means to mask operations as well.

Maj Atkins noted: *Before coming to MAWTS, I never looked at the question of electromagnetic spectrum whatsoever. Now it is a central consideration of my focus and effort.*

In short, the Marines at MAWTS have been working new ways to do FARPs as a way to do EABOs, but there are key limitations to what one can do in the real world.

- And ultimately, the key combat question can be put simply: What combat effect can you create with an EABO?
- How does the joint force use an EABO in creating a joint effect?
- And what is the relationship of the creation of EABOs to what the Marines do when the National Command Authority calls on them to deploy?

Refocusing the Force: MAWTS-1 Works on Ways Ahead

December 14, 2023

During Laird's visit to MAWTS-1 in November 2023, he talked several times with the CO of the command, Col Eric Purcell, about the evolution of the three WTIs he has been in charge of since he took command.

Thinking about a world with multi-polar authoritarian players does tend to focus your mind when you have to train to fight a variety of adversaries. And that is where we started our interview. Col Purcell noted that one of the major changes at the command is focusing attention on the developing capabilities of the kind of adversaries the Marines have to face.

Purcell noted that the Marines who come to the WTIs know learn about three categories of threats: those posed by Chinese forces; those posed by Russian forces; and those posed within the context of contingency operations.

The experience of the land wars and the historical legacy of fighting against Soviet-era equipment does not prepare today's Marines for the conflicts in which they are engaged or likely to have to deal with. It is important to re-shape the curriculum to reflect real world dynamic threats, and Purcell noted that they were focused on doing so.

Col Purcell as seen in the end of course video from WTI-1-24. Credit: USMC

But major challenges face the Marines in doing so, and

MAWTS-1 as the core warfighting training center which standardizes operational preparation needs to adapt its training approach to the evolving threat envelope.

One change which Col Purcell underscored was the need to refocus on tac air and assault support integration. He noted that in the land wars, the two forces tended to operate somewhat in different spheres, but going forward integration was key.

In part, this is due to the range and speed requirements for a Marine Corps insertion force and the opportunity to leverage the advantages which as F-35/Osprey force provides to the USMC along with the arrival of a new generation heavy lift asset, the CH-53K.

But this also due to the changes within the assault force itself, as the Osprey adds new capabilities carried in the back of the aircraft, or its ability to contribute to new missions for the Marines such as ASW support.

A key focus of attention has been moving beyond the training to establish FARP/EABOs, to working on the more important point – what effect are you trying to create through FARP/EABOs.

With current capabilities, the Marines can create EABOs to provide for sensing capabilities, support for the transition of air elements, C2 node creation and support, but until a new generation of weapons show up, limited ability to provide for fires within a transitory EABO.

Col Purcell indicated that a primary focus within FINEX at the end of the course is bringing the different force elements together to test out their capabilities to deal with an integrated scenario.

Two scenarios were notable in our discussion.

- The first was working maritime strike and support. Here the Marines would operate from Camp Pendleton with San Clemente being an enemy location which was being reinforced by "red" combat ships coming from the north. The U.S. Navy has provided ships for this purpose. Tac air provides the main means for interdiction of enemy shipping but simulate longer range strikes from

Pendleton is involved as well. A key capability of fifth generation aircraft is their ability to manage third party targeting which is a capability which the Marines will leverage going forward in terms of tapping into land-based strike, and one might assume this could by Army or Marine Corps ground strike capabilities. Under the strike force, a TRAP force enabled by Ospreys operates and in FINEX, the Marines operated such a force and exercised it with the Navy and the USCG. When the interdiction of the surface fleet threat was attenuated, the Marines shifted their attention to provide ASW support to the Navy, largely by providing sonobuoy deployment support.

- A second key approach will be emphasized in next year's spring WTI. Here the integration of TACAIR and assault support will be the focus of attention against an appropriate scenario for using such a force. With the evolution of tactical air capabilities supplemented with data provided from Reapers and other ISR sources, and the evolution of the payloads carried by the assault force, a variety of scenarios can be tested in the next WTI and future WTIs.

A major challenge though facing the USMC is the dynamic changes within the joint force itself.

- How does the USMC support the joint force if the Navy is in the thrust of significant change sorting through what they mean by distributed maritime operations?
- How does the USMC support the USAF if that force is in the process of sorting out what Agile Combat Employment means in practice?

The fires challenge is a key one in this context. Who is the fires authority in a maritime strike scenario? A USAF air wing? A Navy surface action group?

Until this is sorted out land-based weapons whether operated by the Army or the USMC cannot have the desired effect. And having an effective joint force rests on having the fires authority challenge met and managed.

As Col Purcell concluded with this dynamic area of joint force engagement: "Typically, we refer to Target Engagement Authority abbreviated to TEA.

"In land wars between the Combined Forces Air Component Commander (CFACC) and the Coalition Forces Land Component Command (CFLCC) we typically have this ironed out.

"But it has been such a long time since we have done true Joint Naval engagements that the process for engaging maritime targets and who exactly is the Target Engagement Authority (TEA) or can be the TEA is not as tried and true as it is for land engagements."

8

The Change of Command Ceremony, May 2024

Laird mentioned earlier when discussing his visit to the change of command May 3, 2024 that the interviews he did with two former COs of MAWTS-1 and the current and outgoing CO of MAWTS had significant continuity of thought and purpose about the command and its focus and purpose.

In this section, we include these two interviews, prior to providing a photo album of the ceremony which highlights the tradition and continuity of this impressive and unique command.

ROBBIN LAIRD & EDWARD TIMPERLAKE

Marine Aviation Weapons and Tactics Squadron One
MCAS Yuma
May 3, 2024

Looking Back at the Formation of MAWTS-1 and Shaping a Way Ahead

May 7, 2024

Laird had a chance to talk with LtCol Howard DeCastro, the first CO of MAWTS-1 and LtGen Barry Knutson, the eighth commander of MAWTS-1, the day before the change of command ceremony at MAWTS-1.

We talked about the approach of MAWTS-1 from the beginning and the importance of continuing the tradition and approach going forward for the USMC to operate effectively in today's conflicts and combat situations.

DeCastro started with this comment: "We told people from the beginning, MAWTS does not fight wars.

"We are here to make people as good as they can be when they go to war.

MAWTS-1

"It is the squadron and the fleet marine force that is doing the work.

"At the beginning, some wanted MAWTS to have distinct uniforms and I said that was a bad idea.

"We are part of the force, but just focused on making them better.

"We are training the trainers who go to the squadrons and proliferate the best combat practices to the force.

"We are Marines.

"We are here to make the USMC better.

"Nothing less and nothing more."

LtGen Fred McCorkle on the left and LtCol DeCastro on the right attending the change of command ceremony at MAWTS-1 May 3, 2024.

LtGen Knutson reinforced this point as follows: "Some brilliant people like Howard developed Project 19 and the idea of training the trainers. The trainers who come out of MAWTS are the training gurus of the squadron in the ops department to lead the training program and every squadron would have one or maybe two WTI graduates.

"Prior to MAWTS, training was divided between East Coast and West Coast Marines receiving different training. MAWTS-1 was established to have uniform training for the Marines.

"With MAWTS, we brought in the professionals, who knew what they were doing, and they were forged into an integrated force.

It was from the beginning a center of excellence for training the trainers.

"The capability has only accelerated over the past thirty years. Now they are doing things we never even dreamed about doing and in all domains.

"The approach is to add on modules of new capabilities over time as the force has evolved. When I was here, I added a loadmaster course for the C-130, and we added a ground based air defense course for the Hawk and IR guys.

Major General Bobby Butcher, the second commander of MAWTS-1 on the left with LtGen Barry Knutson, the eighth commander of MAWTS-1, on the right, attending the change of command ceremony at MAWTS-1 on May 3, 2024.

"Originally, we did not teach air-to-air tactics. To get a WTI patch, fighter pilots would go through Top Gun and MAWTS. But we could not get enough pilots through Top Gun to do so. To deal with that we added an air-to-air course after the WTI course so that the fighter pilots would work on their air-to-air tactics as well."

This was the template created from the beginning at MAWTS-1 and because of the modular structure of training – adding modules to the training regime dependent on need and adding of capabilities – it is a template that has been able to grow into today's variant of

MAWTS-1 and also explains why there is clear continuity from its founding until today.

LtGen Knutson characterized MAWTS as "an operational petrie dish. We go out there and we put a FARP 60 miles out, we have F-35s overhead, we integrate all six functions of aviation, and we look at the logistics required.

"New technology gets inserted into a very, very complex and very realistic scenario, and Marines learn how to use any new gear. It's like a crucible or like a petri dish. It's a brilliant approach."

Because of such a dynamic training template, a discussion with the first and eighth commanders of MAWTS-1 is very similar to the current COs of MAWTS which I then had two hours later on the MCAS Yuma base.

An Update on MAWTS-1: The Perspective of the Outgoing and Incoming Commanding Officers, May 2024

May 15, 2024

On the day before the change of command at MAWTS-1, I had a chance to talk with the outgoing commander, Col Purcell and the incoming commander, Col Smith. I had last been to MAWTS-1 in November of last year, where we focused on the WTI or the Weapons and Tactics Instructor Course just completed before my visit.

This time the first WTI course of this year had just been completed and we had a chance to discuss the FINEX event which closes out the course.

Col Purcell started by underscoring that the grounding of the Ospreys by the services after the accident last year with an Air Force Osprey, created a challenge for them. Not having Ospreys – which frankly are a bedrock platform in the transformation of their concept of operations – caused a problem in the WTI. There were some missions they simply could not do, and shifted assets around to do missions which was not their primary mission focus.

Col Smith and Col Purcell at the Change of Command Ceremony, May 3, 2024. Credit: MAWTS-1

When I worked for the SECAF, he had a "day without space" for the USAF, which was not a pleasant experience for the airmen. Similarly, "weeks without an Osprey" was a similar experience, and reinforced how crucial the platform is for the kind of con-ops the Marines have become used to being able to do with an Osprey-enabled force.

Col Purcell talked about the changes that have occurred since taking command. He underscored that one major change has been working in maritime strike packages into the force as well as enabling the ability to do EABOs or Expeditionary Advanced Base Operations.

But he made it clear that EABOs are not an end of themselves: what combat purpose do they meet and how do they make for a more effective force in particular missions?

This is how he put it: "The ability to conduct expeditionary advanced bases, that's a capability that's going to enable something else. It is not a mission of itself. EABOs are what we do in an operational area to project lethality and to project our power and delivering capability to deter an enemy. It has to be about the ability to integrate all six functions of marine aviation in support of a larger mission."

MAWTS-1

With regard to shaping a maritime strike package, Purcell described the journey during his time in command. They went from starting the effort with little involvement of the Navy, to the latest FINEX where the Navy fully participated, and the Marines were able to work a complex strike mission in support of the maritime force.

One mission which has been identified and which MAWTS-1 has been training for is the TRAP mission associated with a maritime strike mission. The need to recover rapidly any personnel downed in a maritime assault mission is something the Osprey is uniquely positioned to do. Only you can't do it if it is not there. Fortunately, the ban on Osprey use was lifted in time for them to be able to use the Osprey in the maritime strike event within FINEX.

The CH-53K has come to MAWTS-1 and in this event operated with CH-53Es and UH-60s which were provided by the Alaska National Guard. They worked the assault package for the course. And in FINEX were part of a complex combined arms assault to seize any enemy airfield against Red Air and ground forces and inserted the ground combat element to seize the airfield. According to Purcell: "I am a huge fan of the CH-53K. It is incredibly capable platform."

Col Smith has been at MAWTS-1 and has been involved with its activities and is certainly ready to take over. As an Osprey pilot, I am sure he is more than ready to continue the growth path of that aircraft in its multi-mission approach being enabled by roll on and roll of capability.

Col Smith noted that a decade ago, MAWTS-1 had already begun to focus on the impact of dealing with a peer competitor in littoral operations. And he underscored the continued growth in Marine Corps capability to operate in such environments, which had been taken forward under Col Purcell's command would be continued under his as well.

As Col Smith underscored: "We go to war twice a year at MAWTS-1. We focus on the evolution of the combat force, and notably, are focused on enhancing our ability to operate against a peer competitor in the littorals. That challenge was envisioned ten years ago here at MAWTS and it is very rewarding for me to come back now and see the progress we have made. We want to take that momentum and carry it forward."

9

Retrospectives

We have interviewed several of the past commanders of MAWTS-1 for this book in order to gain a historical perspective on the founding and evolution of MAWTS-1 and to honor the work of the commanders of MAWTS-1.

In the appendix, we provide a list of all of those who served as commander of MAWTS-1.

The interviews presented reflect the insights of the Marine Corps officers involved with MAWTS-1 during their time of engagement. We indicate the date of the original publication with each interview.

Recollections on the Establishment of MAWTS-1

December 6, 2023

By Howard DeCastro, LtCol USMC, (Ret.)

Trying to describe how MAWTS-1 and the WTI training concept began is very much like the classic tale of the blind men describing an elephant. After Vietnam, there were lots of Marines who were thinking about ways to improve Marine Corps Aviation and there were many initiatives throughout the Marine

Corps. Everyone who participated in the establishment of MAWTS-1 and the development of the Weapons and Tactics (WTI) concept has a slightly different story.

This article presents my recollections.

In 1976 I was stationed at NAS Miramar, scheduled to be the Executive Officer of the first Marine Corps F-14 Squadron. When the Marine Corps decided the F-14 was the wrong fighter for the Marine Corps and canceled its participation, Colonel Bob Norton, Commanding Officer of MCCRTG-10, asked me to come to MCCRTG-10 in Yuma, Arizona.

After closing down the Marine Command at MCAS Miramar, I transferred to MCAS Yuma with several of the Pilots and Radar Intercept Officers who had been part of the F-14 Program and a few of us went to work on building an F-4 Training Program using the Instructional Systems Development (ISD) construct that was used to develop the F-14 Classroom Training System.

Captains Rick Scivicque, Rob Savio, Jim Hollopeter, Jerry Cross and I put together an F-4 ISD based training system with funding from Jim Bolwerk, a retired Naval Aviator, who was heading up the Navy's Training Directorate at COMNAVAIRPAC. The funding went through the Aviation Training Department at Headquarters Marine Corps and was supported by Colonel Paul Boozman. Colonel Boozman would have preferred development of a Helicopter Training Program, but the money was controlled by COMNAVAIRPAC and was earmarked for Navy and Marine Corps F-4 use.

Colonel Boozman knew that Marine Corps helicopter training was in need of a more standardized, focused, and formal program that would teach the tactics and skills necessary for force projection in combat. I emphasize this point because his strong position on the need for improved tactical training for the Helicopter community helped ensure the inclusion of helicopter training and, broadly, inclusion of training for all aviation assets in the development of the MAWTS/WTI concept.

While working on the F-4 ISD project Colonel Norton and his XO Colonel John Hudson made us aware of Project 19, one of the many recommendations on ways to improve Marine Corps Combat Readiness that then Colonel John Cox presented to the Commandant. Colonel Norton directed us to put together a concept of how Marine Corps Aviation training could be improved.

Over the next year, working with people like Bill Bauer (CO of VMFAT-101), Bobby Butcher (XO and then CO of VMAT-102), Dave Vest (XO then CO of VMFA-531 and previous head of the F-4 shop in MAWTU-PAC), we put together our ideas and developed the MAWTS/WTI concept.

Collectively, we had not been satisfied with the application of air power in Vietnam, our only war experience. There were numerous tactical applications that were effective (A-6 strikes in the north; close air support; helicopter troop insert, withdrawal, and rescue; C-130 tactical resupply; and breaking the siege of Khe San are examples).

However, an overall strategy, the integration of aviation resources, and effective coordination with the ground elements was less than ideal. It was clear we lost more Marines than we should have because we were not well coordinated with the Ground Forces and weren't well coordinated with other squadrons, especially with other types of fixed wing, helicopter, and transport aircraft squadrons.

In the 1960s and 1970s, Fleet Squadron training depended on the quality of the more experienced pilots and whether they would teach what they knew or just try to beat you and hope you would learn.

The Special Weapons Training Units and later MAWTUPAC and MAWTULANT had good Attack and Fighter training and certification programs that helped the A-6, A-4 and F-4 communities standardize and improve their training and combat capability, but these were generally specific to the aircraft type and didn't integrate the full range of aviation types and capabilities.

The Navy's "Top Gun" school that started in the 1970's was excellent for the improvement of individual Air Combat Maneuering (ACM) skills but was available only to F-4 fighter crews.

Integrated training with the Marine Corps Ground Forces was

also less than optimum. While there were opportunities to participate in some Ground Training Scenarios, the lack of participation in planning and debriefing led to exercises that failed to build the kind of integrated support that is necessary to be highly effective in combat.

We developed a concept that would include all aviation assets, working together as a coordinated and integrated team, to support a Ground Scheme of Maneuver.

We fully understood that the value of the Marine Corps was invested in the troops on the ground who could defeat the enemy and take and occupy the terrain that is critical for winning battles and wars and knew it was our job to provide the best possible air support to enhance their combat capability.

While we were working on the concept at MCAS Yuma and El Toro, California, there were other Marines throughout the Corps who were developing similar and alternative concepts. It was clear the Marine Corps was going to do something significant to improve the training and application of our aviation assets. It remained to know where, what, and when.

Key elements of our concept were:

- The inclusion and integration of all Marine Corps aviation assets, every type of tactical aircraft, command and control, logistics, and air defense.
- Each year, train one pilot/aircrew from each squadron in the Marine Corps as Weapons and Tactics Instructors (WTIs) who would act as instructors and operations coordinators in their squadrons. Assign those pilots/aircrew back to their squadron for three, or at least a minimum of two years so that, once established, every squadron would have two WTI pilots/aircrews.
- Provide ground training on individual aircraft and the broad range of Marine Corps aviation assets to learn how to integrate and utilize the capabilities 0f the full range of Marine Aviation to support the Marine Corps Ground forces.

MAWTS-1

- Provide individual and combined-force flight training utilizing the best-known tactics for air combat and defensive combat maneuvering, ground attack, close air support, troop insertion, logistics supply, electronic warfare, command-and-control, and reconnaissance.
- Provide instruction on the capabilities of our known and potential enemies and the best tactics to defeat them.

Some senior officers thought the Command of MAWTS should be restricted to Fighter Pilots. We thought it important to share command of MAWTS among the communities to ensure broad buy-in of the concept and recommended that the Commanding Officer shift from fixed wing to helicopter aviators every other change of command.

We also recommended that the incoming Commanding Officer should serve one year as Executive Officer to ensure a smooth transition from Commanding Officer to Commanding Officer, like the "Fleet-up" construct used in Navy Squadrons.

We didn't alternate Command immediately because we had not brought a Helicopter pilot in as Executive Officer and thought it important that the next Commanding Officer be someone who was thoroughly familiar with the background and goals of MAWTS/WTI.

I recommended that Bobby Butcher be my relief because he was a proven leader, had helped develop the concept, fully understood what we were trying to do, had a successful Squadron Command, and could easily take over and maintain the momentum we had established.

Colonel Butcher cemented the practice of rotating the Command between fixed and rotary wing aviators and the Fleet-up construct by selecting Jake Vermilyea, a transport and helicopter pilot as his Executive Officer and eventual replacement as the third Commanding Officer of MAWTS-1. Colonel Butcher did this even though there was significant push-back from some Senior Fighter Pilots. It helped that LtGen White, a Helicopter pilot, was the

Deputy Commandant for Aviation and supported Colonel Butcher's position.

It is an understatement to say that not all Marine Aviators supported the MAWTS/WTI concept. Many thought it was far too expensive. There was significant push back on the inclusion of helicopters and transports based on the fear that the current MAWTU programs supporting A-6, A-4, and F-4 training would be watered down.

There also was concern about Yuma as the location. Notably, Colonel John Ditto, Legislative Assistant to the Commandant, who had been the head of the F-4 Fighter Shop in MAWTULANT and was familiar with the ranges on the east coast, thought the new unit should be stationed out of Beaufort or Cherry Point.

The concept we put together in Yuma was pitched at a conference held at El Toro, chaired by Colonel Don Gillam of FMFPAC. Although some of the conference attendees did not fully support the concept as presented, Colonel Gillam deemed the conference enough of a success that he sent LtGen Andy O'Donnel, Commanding General FMFPAC, a message that FMFPAC should support the concept.

As a result of LtGen O'Donnel's support, LtGen Tom Miller, Deputy Commandant for Aviation, asked for a briefing. LtCol Bill Cooper who was at HQMC AAP, presented the brief that we had developed and LtGen Miller approved the concept.

Concurrently, at Headquarters Marine Corps, a great deal of thought and planning had gone into the development of improved Aviation Training. Parallel with the development of our MAWTS/WTI concept, there was an initiative to merge the two MAWTUs.

The HQMC initiative and the MAWTS/WTI concept fit well together. A decision was made to have the new unit report to Aviation Training, where Colonel Paul Boozman was influential. His vision, direction, and support were important in forming MAWTS-1, from gaining support from the Commandant, providing the funding, approving the manning, and choosing the location.

It was decided to conduct a trial course under the direction of

MAWTS-1

the MAWTUs. MAWTUPAC under LtCol Ray Hanle was selected to lead the effort. Instructors from the two MAWTUs, augmented by other highly respected pilots and crew members presented the first Weapons and Tactics Instructor (WTI) course in 1977.

The course was enough of a success that a second trial class was conducted in early 1978. The success of those two WTI classes led to establishment and Commissioning of Marine Aviation Weapons and Tactics Squadron-One (MAWTS-1) in June of 1978 with 35 officers and 19 enlisted Marines.

In researching this article, I have been reminded that there was a trial course held by MAWTULANT on the East Coast. That course apparently was not as well received as those held at Yuma, which helped cement the decision for Yuma as the best location.

A great deal of credit for the successful initiation of the MAWTS/WTI concept lies with Colonel Hanle and the MAWTU instructors who conducted those two first trial courses.

Much of the early success of MAWTS-1 came from the steady support of Colonel Boozman at Headquarters Marine Corps Aviation Training, and from LtCol Duane Wills in the Officer Assignment Branch.

Colonel Boozman was instrumental in shaping the concept, and supporting and approving the training and command constructs, ensuring MAWTS-1 was fully staffed, and that it was adequately funded.

During my two years with MAWTS-1 we conducted four WTI Courses. Support from the Air Wings was excellent, and support from VMAT-102 and VMFAT-101 was outstanding, providing maintenance support and even aircraft when required.

There are some interesting stories about those first two years, such as the MAWTS-1 CH-53 Instructors being pulled out to support the training for the Iranian Hostage Rescue, a trip to Israel to confer with the leaders of the Entebbe Rescue, a Final Exercise evolution that was conducted in marginal weather, and the unfortunate loss of a CH-53 with Crew and passengers and loss of an F-4 Crew.

Near the end of my tenure, General Wilson, Commandant of

the Marine Corps, visited the command for a briefing. In that brief I raised some issues of concern about the mobility of Command and Control, lack of armament for helicopters, and paucity of anti-air capability. While this was not the brief General Wilson had expected, it led to MAWTS being tied in with the Deputy for Aviation for discussion, participation, and influence on a variety of aviation issues.

I have had the pleasure of visiting MAWTS-1 several times since I retired in 1980. It appears that every Commanding Officer, with the support of the excellent MAWTS-1 Staff, has improved its value to the Marine Corps. The direct involvement of Secretary Lehman helped bring MAWTS-1 training to a higher level. MAWTS-1 has grown in size, mission depth, and overall capability.

We thought, when it first began, that Marine Corps Aviation would be improved and that someday most, if not all, Wing Commanders would be former MAWTS Instructors and/or WTIs which would certainly improve the coordination, integration, and combat capabilities of Marine Air. That has happened. The Marine Corps has even had a Commandant who was a MAWTS-1 Instructor.

Note: The author of this article was the first commander of MAWTS-1.

Col Paul Boozman and the Creation of MAWTS-1

December 6, 2023

Col Boozman has had an extensive Marine Corps combat career. With his time in Vietnam, he developed the conviction that the Marines needed to make a significant commitment to training and preparation for the Marines who also face the certainty of deployment.

As the video which highlights the most recent FINEX for MAWS-1 starts: "It is not really a matter of if the Marine Corps goes it to combat, it is a question of when."

But now these Marines have the advantage of the training

MAWTS-1

provided through MAWTS-1. Col Boozman and his generation of Marines did not.

Boozman and the founders of MAWTS-1 wanted precisely to carve out a path whereby today's Marines have a significant training advantage which they did not have.

We had the honor and privilege to talk to Col Boozman and the day of his 90th birthday, November 15, 2023. It was an experience we will never forget, being in the presence of a Marine who made a difference for the generations that followed.

Col Boozman in Vietnam.

We started the interview by discussing his career and his path to be a founder of MAWTS-1. No war games here: just learning forged in combat experience.

Col Boozman: *I have always been a Marine Fighter aviator. I got my commission in 1954, finished flight school and then went to Cherry Point where I flew the F-9F-5 in a fighter Training Squadron.*

After fighter training, I was transferred to Japan to begin a career of 20 years which included three tours in Japan and one year in Viet Nam, flying over 200 combat missions. During these years, I flew all models of the FJ, the F3D, F4D, F8U and the F4H Phantom 2 (models B, J and N).

Additional assignments were duties with the 2d Marine Division as Air Liaison Officer and Forward Air Controller which took him to Panama and Turkey for training exercises. During his career he attended the following service schools:

- Forward Air Controller School
- School of Naval Justice
- Marine Corps Communications Officer School
- Armed Forces Staff College

His formal education was a Bachelor of Arts in Mathematics and a Master of Arts in educational technology.

His command experience included Commanding Officer, Marine Air Control Squadron 6, Executive Officer Fighter Attack Squadron 323 in Vietnam, Officer in Charge of the Airfield, Khe Sanh Combat Base, Executive Officer of Air Reserve Training Detachment, Dallas, Executive Officer and Commanding Officer of Marine Aircraft Group 15 in Japan.

When Boozman then came from MAG-15 in Japan to headquarters Marine Corps where he was attached to the Aviation Department.

Headquarters USMC was reorganized at the time with all training activities were then placed under the control of the Operations and Training staff group. Training activities were divided into ground and into aviation training, with Boozman placed in charge of air training within the Operations and Training Directorate of Headquarters of the USMC.

Boozman laid out in detail how the pilots and Marines generally were making it up as they went along in Vietnam. And there was no standardization whatsoever built into the combat learning process.

He underscored that this experience was seared into him, and he was committed to not having the next generation of Marines to have to suffer from the lack of training which reflected the reality of the combat they would face with the weapons they possessed and the nature of the forces they would work with.

Rather than playing pick-up sticks, why not build a modern training approach which would build comprehensive knowledge standardized across the force?

A commitment to doing this would be the driving force behind Boozman and his colleagues' focus on what a new training regime needed to look like.

Boozman then organized a group at headquarters to shape ideas for a new way ahead. He insisted on the following: *We are never going to repeat how we fought in Vietnam. We want to make sure that people know what they are doing. We want to come up with a way of training where people know what they are doing when they go into combat.*

Boozman and his team focused on an approach whereby the schoolhouse would train the trainers who would go back into the units, shape standardization at the schoolhouse, take it to the squadrons and then the combat experience would be cycled back into the schoolhouse training and create a continues learning process which shaped the training experiences.

LtGen John Miller signed off on the concept and then they presented the plan to LtGen Tom Miller, who had to supply the personnel to staff to staff MAWTS-1.

According to Boozman: *LtGen Tom Miller was hoping to be able to reduce the number of persons committed to training, but after being briefed on the concept and scope of reorganization of the training program, he approved the personnel requested. That was the original foundation of MAWTS-1.*

And there seemed to be resistance to the idea of the MAWTS concept. The Commandant at the time was General Louis Hugh Wilson, Jr., a World War II recipient of the Medal of Honor and 26th Commandant of the Marine Corps and he gave his full support to the MAWTS proposal and MAWTS was commissioned under the direct supervision of Aviation Training.

At Colonel Boozman's request, LtCol Howard DeCastro, a key player in the development of the MAWTS concept, was appointed as its first Commanding Officer.

In 1980, after several MAWTS classes, Colonel Boozman. retired from active duty. He was awarded the Legion of Merit Medal for his work in aviation training.

The Perspective of MajGen John Cox: Present at the Creation of MAWTS-1

December 18, 2023

MajGen John Cox entered the Marine Corps via the Officer Candidate program at Quantico, Virginia, and was commissioned a Marine second lieutenant in September 1952. After completing the Basic School, he reported for flight training and was designated a Naval Aviator in July 1954.

Successive assignments were with all three active Marine Aircraft Wings and with the 4th Marine Aircraft Wing: service on the USS Lake Champlain; duty as an air and naval gunfire platoon commander; duty with Marine Fighter Squadron (VMF) 451, VMF 333, and VMA 324. He completed Communications Officers School, served as Communications Officer at MAG 15, El Toro, and later was Operations Officer of VMFA 513 at El Toro, Atsugi and DaNang (1962 1965).

He graduated from the Armed Forces Staff College in 1967, and after serving as an instructor at the Naval War College, he returned to Vietnam for a second tour of duty, serving as Commanding Officer of VMFA 115 and as Executive officer of MAG 13 at ChuLai.

Tours of duty in the 1970s include Executive Officer of MCAS, Kaneohe Bay, and duty on the staff of the Commander in Chief, Pacific. After graduation from the National War College in 1974, General Cox reported for duty at Headquarters Marine Corps. He was promoted to brigadier general on Nov. 4, 1977.

Then he served as Assistant Wing Commander, 1st Marine Aircraft Wing, Commanding General, 9th Marine Amphibious Brigade, Assistant Chief of Staff at Headquarters Marine Corps, Deputy Chief of Staff for Research, Development and Studies at Headquarters Marine Corps, and Commanding General, MCAS El Toro/ COMCABWEST, followed.

He was promoted to major general on April 9, 1981, with a date of rank of Aug. 1, 1978. He assumed command of the 3d Marine Aircraft Wing, Marine Corps Air Station, El Toro, CA, in May

1981. In June 1982, he was assigned duty as the Director for Operations, J 3, for the Commander in Chief, Camp H.M. Smith, Hawaii. He served in this capacity until July 1, 1985, when he retired from the Marine Corps on July 1, 1985.

We had chance to talk with him on December 4, 2023 to get his perspective on MAWTS-1.

During his career. MajGen John Cox amassed 5,043 hours total flight time and was a rated pilot in nearly every Marine Corps fixed-wing aircraft of the Cold War era, including the F6F Hellcat, AD-4 Skyraider, FJ-4 Fury, F8U Crusader, and F-4 Phantom II.

MajGen John Cox provided important insights with regard to the origin of MAWTS. When he came out of the National War College in 1974, he went to work at Headquarters USMC in Aviation where he was for the four critical years which would birth Project 19 and eventually MAWTS. He underscored that a key person in all of this was Lt. Gen. Philip Shutler, one of the pioneers of modern USMC aviation.

According to Cox: *Shutler as considered the idea man for USMC Avia-*

tion, and he did a good job of pulling things together and coming up with achievable and important missions.

While working with Shutler, they came up with 21 projects to improve USMC aviation. Project 19 was focused on training and became the context for the creation of MAWTS-1. Cox was there at the creation, and we can thank him for his contribution to the USMC and to the nation.

Institutionalizing Excellence: The Raison d'être for Creating MAWTS-1

November 28, 2023

We had a chance to look back at the creation of MAWTS-1 with Col James Davis a Vietnam veteran and one of the team members who created MAWTS-1.

Davis was commissioned in 1963 and for his tour in Vietnam served as an infantry officer in the First Marine Brigade in 1965-1966, Platoon Commander, 2nd Bn, 4th Marines. When he returned to the states, he went to flight school.

There he became a Phantom F-4 pilot. He would fly 326 combat missions in the F-4 in Vietnam. He was a member of two squadrons that served in Vietnam, namely VMFA-323 and VMFA-314. In 1972 he was based at Nam Phong, Thailand referred to as the "Rose Garden." His service dates were from 1963 through 1988.

When he returned from Southeast Asia, he did two tours at MCAS Beaufort Air Station and from there went to Headquarters Marine Corps to work in the Aviation Department. When he arrived the project to create MAWTS was being shaped.

Before MAWTS, there was an East Coast training facility at Cherry Point which was a ground school with no flying and a West Coast training facility at Yuma. But the training was very different and not delivering standardized preparation for operational conditions.

MAWTS-1

Photo taken in Chu Lai RVN 1969 with fellow Marines.

According to Davis: *The JOs in the Aviation Department were fully behind MAWTS. We did not wish to repeat our Vietnam experience which was learning by the seat of our pants.*

There was significant opposition at Headquarters to creating MAWTS, but we believed in MAWTS because we thought it was a wonderful idea to have a location with standardized training worked through a big exercise.

We worked a briefing for the head of Aviation, General Miller, and we worked it hard because we all believed in MAWTS. At the meeting which would make the decision to launch MAWTS, all the senior people were there, and it was clear the Commandant wanted to do this.

General Wilson was a formidable leader. He controlled the conversation and said we are going to do this. Did anyone object? Well, they did not and MAWTS was launched and came to life.

The importance of MAWTS for Davis was simply put: it was the standardization of training.

As Davis put it forcefully: *The purpose of MAWTS is to institutionalize excellence. End of story. Period.*

The Perspective of the First Commander of MAWTS 1: LtCol Howard DeCastro

December 14, 2023

DeCastro had been kind enough to provide an overview paper on the founding of MAWTS and its first years. This allowed us in our interview with him to highlight key takeaways as he went into the job of being the first commander of this new organization.

The first takeaway was that the Vietnam War generation of aviators in the USMC had a unity of purpose, at least those associated with the creation of MAWTS. They did not have standardized training and had to learn on the job and did NOT have the kind of coordinated operations that would make operations more effective.

As DeCastro commented: *There were F-4 RIOs, A-6 BNs, Helicopter Crewmen, and others associated with combat in Vietnam who, like many pilots, were dissatisfied with our performance.*

The second takeaway was that he came to MAWTS while working almost full time on Project 19 which was the birthing of MAWTs. By working with the team doing Project 19, when he went to MAWTS he had a clear idea of what to do.

This is how DeCastro combined the two takeaways. *We were taking resources that existed and simply did not work well together and did not have good training to do so. We thought we could make the USMC better by getting everyone on the same page and train together the way you would go to combat and see what problems arise during our training and address them.*

He underscored that it has been important for MAWTS to continue "to keep command excellence" going forward. We discussed the tradition which started early of rotating commands between a fixed wing and rotorcraft/tiltrotor craft pilot as being important to forge unity of purpose in the training process.

DeCastro also highlighted that the command brought into the first WTIs very competent Marines. These Marines then are trained and go back to their squadrons to work the integration challenge going forward.

What is very notable is that what DeCastro and his mates put in place and in motion in 1978 is not only still going, but it is getting better. But it follows the same trajectory set in motion more than 40 years ago. Rather amazing when you realize that.

The Second Commander of MAWTS-1: Bob Butcher

November 27, 2023

We had the opportunity to talk with Major General Bobby Butcher on November 16, 2023, to learn how he worked the challenge of shaping the institutional framework for the kind of standardized advanced training which, after all, was the whole purpose for creating MAWTS-1 in the first place.

Butcher in his flight suit at Chu Lai, South Vietnam, 1967.

But setting a framework that has not been created requires significant creativity, imagination, bargaining and hard work. It was clearly a team effort within which Col Butcher worked with several MAWTS founders to set in motion the trajectory on which MAWTS was launched.

Major General Butcher had a very distinguished career which

can be seen in the document we have included at the end of the article.

Here we are focusing more narrowly on how an experienced A-4 pilot came to be the second commander of MAWTS-1 and how he networked with Marine Corps leadership to carve out a way ahead which has persisted to this day.

Butcher highlighted several aspects of the challenge of establishing and then building a path forward for the kind of vision which the founders of MAWTS-1 clearly had.

- The first aspect was the question of working MAWTS-1 WTIs to include both the best rotary wing as well as fixed wing pilots and crews.
- This was worked by Butcher on two levels: who would succeed him as at MAWTS? And how to get the best helo pilots and crews participating in the WTIs.
- The second aspect was working or struggling to ensure that all of the key aircraft flown by the USMC participated in WTIs so that the full gamut of warfighting capabilities could be exercised at MAWTS and then taken back to the squadrons by the WTI instructors.
- The third was staying in touch with the squadron commanders and air wings world-wide and providing input as well as receiving their insights with regard to changing operational conditions and folding that into the ongoing MAWTS training experience.
- The fourth aspect was working with and learning from allies during WTIs and visits to Yuma. To shape combat excellence, one should not limit oneself to what the U.S. forces alone are doing. Combat learning from allies is a key part of the training process.

Let us look at each of these aspects as discussed with Butcher. With regard to the rotary wing issue, Butcher highlighted this in two ways – the challenge of getting the properly qualified pilots and working his replacement as CO of MAWTs.

Butcher noted:

We were doing well with regard to our first WTI when I was in charge, but the only area we were not doing well as with regard to helicopter pilots. The list of candidates sent to us made it clear that the pilots who were selected to come simply were not qualified.

I could not get through to these helo COs because they would say "Butcher, you don't understand the trouble we have trying to train, when the division is always asking for all the assets we have available."

I called my boss, the head of training in the Marine Corps, and indicated the problem. I then flew back east to meet with him and then met with General William White who was head of aviation at headquarter Marine Corps. White was a helicopter pilot who certainly saw the need for what I was concerned about.

While I was there, I was given more than 70 files to review on lieutenant colonel helo pilots scheduled to move during the summer of 1981.

Originally, three records were held out from my review, and when I got these, I realized that any of them could be good for my replacement, but Jake Vermilyea was in Washington attending the National War College and I could explain MAWTS-1 to him. I invited him to visit me at my hotel and we met, and he was enthusiastic to come as my XO and then to replace me.

Thus, begun the tradition of rotating command between a fixed wing and rotary wing commander to head MAWTS.

The second aspect was highlighted in story which Butcher recounted of his dealings with 2nd MAW when he needed EW and C2 aircraft to support his WTI. He was told by the MAW commander that they were holding an exercise at the same time and could not provide the aircraft and crews to MAWTS.

But only 2nd MAW had the aircraft needed so Butcher let the 2nd MAW commander know that he was going to appeal to General White to get those aircraft but only after the 2nd MAW commander had his call with DCA.

He did so and DCA saw the importance of bringing those specialized assets to the WTI, for it was MAWTS that was working standardized training for the entire Marine Corps not just a Wing.

Third, while CO of MAWTS-1, Butcher set up communications with Marine Corps operational leaders worldwide.

According to Butcher: *You could call me from any place in the world and reach my desk. I think the extension was 2056 I don't know why I remember that.*

You could call me from anywhere in the world at the Yuma prefix and that number and it would ring on my desk. and I gave that number to all the group all the squadron COs and it was not unusual for me to get calls from them for various things.

Butcher was CO of MAWTS-1 from 27 June 1980-5 August 1982. This meant during his time there, the Israelis had engaged in the conflict in Lebanon. We then discussed the IDF and working with the IDF to shape lessons learned.

As Butcher noted: *The first time the Israelis came when I was at MAWTS, they told me that we were doing the most realistic combat training in the world. We had much to share then as now with the IDF.*

In short, Major General Butcher carved out a beginning orientation which helped shape the trajectory of change which MAWTS-1 continues until today.

The Third Commander of MAWTS-1: Major General C.L. Vermilyea

December 20, 2023

Clyde L. "Jake" Vermilyea enlisted in the Marine Corps in 1955 and served in all three Marine Divisions as a Naval Gunfire Spotter and Radio Operator, later becoming an Airborne Radio Operator in 1959. In 1960 he began flight training as a Marine Aviation Cadet (MARCAD) and was commissioned a second lieutenant in 1961.

He served in all three active Marine Aircraft Wings as a C-130 pilot, transitioned to helicopters (CH-46s) in 1969 (HMM-164 Viet Nam) and flew both fixed and rotary wing until retiring in 1993 as a Major General.

Vermilyea served as the third commander of MAWTS-1, and the first helicopter pilot, beginning the practice of rotating the command of MAWTS-1 between a fixed wing and helicopter/tiltrotor pilot that has served the Marine Corps well, and continues to this day.

We had the privilege to talk with Vermilyea on Dec. 5, 2023 to get his perspective on the founding of MAWTS-1 and the projected long term benefits to the Marine Corps and, in fact, the national defense, of MAWTS-1 and the WTI concept.

One of our key takeaways from the discussion with "Big Jake," as our other interviewees referred to him as, was the background he brought to the command. He had served in two different aviation communities before coming to MAWTS-1, neither of which had had the benefit of the tactical fixed wing training vehicles such as MAWTULANT/PAC, the predecessors to MAWTS-1.

His experience was that although the transport and helicopter communities had in-community training methods, they had not been fully integrated into the combined arms concept central to the Marine Corps' Marine Air Ground Task Force (MAGTF) approach to warfighting.

He had been involved in Project 19, the precursor to the standing up of MAWTS-1 and enthusiastically supported the concept.

The WTI program would infuse the full integration of Marine Aviation knowledge and training throughout the Marine Corps. No matter your MOS, you knew what the other guy was doing. This highlighted the importance of the Tactical Air Command Center (TACC) being deployed to the WTI course and incorporated in the syllabus and exercises. It also ensured proper use and integration of the all-important communications necessary in combined operations.

Thus, the individuals with varying backgrounds contributed in different ways. Their unity of purpose was animated by the different personalities and skill sets of the founding generation of MAWTS-1, but through this process a template for training for what LtGen Rudder has called the "physics of combat" which allowed for innovation going forward.

MajGen Vermilyea stated that while MAWTS-1 is an aviation

unit, and works closely with the Aviation Department, it is sponsored by, and answerable to, the Training Department of HQMC, almost always under a ground forces general. While this setup has worked well it has the potential for complications.

During Vermilyea's tenure as CO, LtGen William Fitch became the Deputy Chief of Staff, Air (DCSAIR). LtGen Fitch had a jaundiced view of MAWTS. During his courtesy call to the new DCSAir, LtGen Fitch informed Vermilyea that one of the first things he intended to do was to shut MAWTS down.

Vermilyea's response, after recovering, was to request the general visit MAWTS during a class, to include the flight phase, and to make the judgement based on his own observation rather than other people's opinion. LtGen Fitch agreed to do that, and in fact made several visits, including flying the A-6 in training and even in the final exercise.

At the end of that episode Fitch told Vermilyea he had "the best squadron in the Marine Corps." Fitch's support thereafter was strong and unequivocal.

We asked Vermilyea why he thought Fitch changed his mind?

Vermilyea underscored that Fitch saw that it worked and had enormous potential for the future of aviation.

One aspect Fitch noted was the use of night vision goggles in helicopters doing low level and Nap of the Earth flying, and the potential for fixed wing aircraft. This 'learn while you train,' and plowing that idea into the way ahead for the operating force is a key contribution of MAWTS-1 and the WTI program and how it works.

During his tenure, Vermilyea saw a growing need for liaison with the other services and outside agencies. To not disrupt the work of the instructor staff he created, with the help of LtGen Fitch, the Aviation Development and Tactics Evaluation section (ADT&E) was created which was comprised of instructors who had finished their normal tour but extended a year.

This released them from the many and varied instructor duties and allowed them the freedom to travel and liaise with other services and entities, and work closely with DCSAir, for new ideas and innovations.

Vermilyea closed by underscoring his key mantra for effective leadership is to understand and apply RAA-Responsibility (cannot be delegated), Authority, (which can be delegated), and Accountability. Leave one out and nothing works. At MAWTS, if you broke the Rules of Engagement (ROE) you went home.

And that brings a key final observation. Although the instructors at MAWTS, and the students who attend the classes are selected for their flying skills and leadership potential, the mission is integration of the combat force for the total mission, not individual performance. The MAGTF, with properly trained and outfitted Air, Ground, and Logistics units becomes a lethal force.

The Fourth Commander of MAWTS-1: Randy Brinkley

November 4, 2023

Randy Brinkley, call sign "Dragon," was the fourth commander of MAWTS-1.

As Dragon explained it: *Howard DeCastro was the first commander of MAWTS-1, followed by Bobby Butcher who become a Major General, then Jake Vermilyea who was the first helicopter commanding officer who headed MAWTS-1 and I was number four.*

He explained: *For the fixed wing F-4/F-18 aviators, they would go through Top Gun prior to coming to MAWTS-1, meaning that they would have mastered their air-to-air combat skills before becoming MAWTS-1 instructors and working the USMC focus on support for the ground forces.*

The F-4/F-18 aviators had to first go through a MAWTU Air Combat Tactics Instructor ground/flight syllabus before being considered for attendance to Top Gun. Howard DeCastro was an early graduate of this training and one of the very best F-4 aviators in the Marine Corps. Bobby Butcher was also one of the best A-4 attack aviators in the Corps. Both were most deserving to be early commanders of MAWTS-1!

"Dragon" described the evolution of MAWTS-1 in those early days as upon rapidly shifting from being an aviation training facility to one supporting the MAGTF. He emphasized that John Lehman who was Secretary of the Navy during "Dragon's" tenure as commanding officer of MAWTS-1 provided significant leadership in this transformation and taking the MAWTS-1 template and applying it Navy-wide.

According to "Dragon": *John Lehman made MAWTS-1 a Navy-wide strategic asset that he used as a catalyst for Top Gun to move to NAS Fallon and be interfaced with the newly MAWTS-designed Strike University to support the integration of the carrier air wing training.*

He saw using the MAWTS-1 template as a way to shape the attack and fighter communities to talk to one another and to operate in an enhanced integrated fashion. As a reserve officer, he came to MAWTS-1 and found a home and took that experience with him as a serving Secretary of the Navy.

John Lehman with Randy Brinkley from a photo hanging in the MAWTS-1 bar.

Dragon underscored that the decision was made to alternate commanders from the fighter to the rotary wing community to lead MAWTS-1. Jake Vermilyea was the first rotary wing commander and Dragon served as his XO and when his time came. he became the commander.

Dragon entered the USMC as an infantry Marine and serving

in Vietnam he determined that he wanted to become an aviator for the rest of his career.

As a Company Commander at Khe Sanh he watched an F-4 Phantom drop out of the clouds to lay down "snake and nape" outside the wire and determined he wanted to do that.

He then received orders to flight school from combat: *You could support the Marine on the ground but fly back to your base at the end of the day and have a beer. That seemed to me a better choice. I would much rather go in harms way than order young Marines to do so.*

"Dragon"'s XO was the legendary Fred McCorkle, who was a CH-46 officer. As Deputy Commandant for Aviation, then LtGen McCorkle would prepare the way for the Osprey and the F-35 which truly has put the USMC in the leading position for air-enabled distributed operations.

Throughout our discussion, Dragon underscored how MAWTS-1 is focused on the practical "doing" of integrated force operations. At one point in the discussion, he put it this way: *We are focused on how we integrate the force, the MAGTF. We could do things that nobody's ever thought of because that's what Marines do. We're thinking outside of the box. That is what Marines have always done, have to do. But in MAWTS-1 you are in an environment that allows you to do it on steroids.*

"Easy" Timperlake then asked "Dragon" to talk about how the Israeli Kfirs become the foundation for adversary squadrons operating with the USMC and the US Navy. This story also provided insight into Lehman's key role in the evolution of MAWTS-1.

This is how "Dragon" described the situation: *One Sunday afternoon, prior to Lehman flying his reserve aircraft back to Washington DC from Yuma, he asked: 'Is there anything I can do for you?'*

"Dragon" answered: *"Well we need aggressor aircraft to work with us during our WTIs. We are having difficulty convincing Top Gun to come over and play that role."*

The next day I received a call from Lehman where he informed me that he talked with the Israeli Prime Minister, and they were going to provide three squadrons of Kfirs to play the aggressor role at MAWTs-1 and with the Navy as well.

Lehman indicated that the Navy and Marine Reserve pilots would operate the aircraft and the Israelis (IAI) will have a maintenance contract to support the Kfir aircraft flight operations.

I was speechless but not for long because my next call was from LtGen Keith Smith, head of Marine Aviation who ordered me to meet with him at 0900 the next day in his office for a meeting with the Commandant at 10 to explain why this new program landed in his budget and how and why did this happen. The Commandant put it bluntly to me: "Brinkley you better make this damn thing work!

"Easy" then asked about the ranges available to MAWTS-1 and their importance. Dragon underscored that why MAWTS-1 is at Yuma is precisely because of the ranges. The Navy ranges, the Air Force ranges, the Yuma Proving Ranges as well as the ranges at MCB 29 Palms are all within the reach of operators from Yuma.

That is why Yuma was the clear choice for integrated aviation training. Brinkley mentioned in passing that when the IAF sent two pilots to the WTI course, they were awed by the training ranges which were larger than the entire country of Israel.

One observation that Laird made was that the kind of integration worked at MAWTS-1 focuses on the distributed force. Now that the U.S. and allied militaries are focused on various forms of force distribution, the Marines are obviously pioneers and continuing to lead in many ways how to practically shape distributed forces.

"Dragon" underscored that from the beginning, the leaders of MAWTS-1 were focused on practicing how to do FARPs. He highlighted that Jake Vermilyea was the founder of this effort and Fred McCorkle was the follow-on catalyst.

- How do you bring the different kinds of aircraft together?
- How do you do the logistics?
- How do you do the C2 for a distributed force in an austere environment?

As "Dragon" put it: *This is in the DNA of the MAWTS-1 community and an incubator for change in the MAGTF and the joint force. It began that way and it continues today with MAWTS-! Being an integral element of the MAGTF training at MCB 29 Palms.*

And incorporating this experience, one can realistically shape requirements and approaches to next generation platforms, systems, and capabilities.

Dragon noted: *There are two critical aspects for successful distribution operations: logistics and command and control. You have to have the operators practice and solve the logistical and C2 problems. You have to figure out how to do so with what you have.*

At MAWTS, the CH-53 guys, the F-35 guys and so on have to figure out how their platform fits into such an operation. And based on this experience, you can think through what future needs can be met to do it better.

Brinkley emphasized that with regard to future developments, the Advanced Development and Tactics Evaluation department at MAWTS-1 continues to play a key role.

These are the warfighters and strategic thinkers like Col. Mike Mott who first headed up ADT&E for MAWTS-1. These are the people that that are the best of the best. They're guys that know what we need in the next generation of platforms to optimize our warfighting capability.

How MAWTS-1 Works on Combat Innovation: A Look Back with Col Michael Kurth

November 30, 2023

We had the privilege to interview former Col Michael "Spot" Kurth on November 15, 2023, to get his perspective on MAWTS-1's early years. Kurth had a distinguished career which included receipt of the Navy Cross for his service in the first Gulf War.

Although we discussed many aspects of his career, our primary focus was on his time with MAWTS-1. What he shared with us were perspectives on how MAWTS-1 provides combat innovation critical to combat success.

He characterized MAWTS-1 leadership as follows: *I would say that Lopp spoke truth to power. In a meeting with the CMC, he challenged the attending Wing Commanders on their readiness postures because they were not addressing training shortfalls and he ultimately prevailed.*

Thunder expanded the reach of MAWTS-1 into the fleet with regular unit visits and expanded relationships with Allied Air Forces. He also focused on expanding rotary wing participation and recruited Big Jake, a rotary wing pilot to be his relief.

This established the precedent of rotating MAWTS-1 command between fixed and rotary wing aviators. "Big Jake" raised the bar for rotary wing by supporting operational requirements in survivability gear, defensive weapons systems for assault helicopters, better night vision devices, defensive air-to-air maneuvers, and large-scale lifts with improved attack helicopter weapons including offensive air-to-air capabilities.

A close working relationship with Task Force-160, the Army's Special Operations helicopter force, brought improved mission planning and night training techniques with their participation in the WTI course.

"Big Jake" and "Dragon" believed that a small cadre of instructors representing the six functions of Marnie Aviation should remain for an additional year in a MAWTS-1 organization where they were free to apply lessons they learned as instructors to larger issues within the Marine Corps.

"Dragon" established a close relationship with Navy Secretary John Lehman (himself a Reserve A-6 BN with some Vietnam experience) who was not happy with Navy Aviation's losses during a carrier strike over Lebanon. Lehman was hugely impressed with the concept of MAWTS-1 and he lent his full support to improving the squadron and extending the concept to the Navy resulting in the establishment of Navy Strike Warfare University.

In addition, Lehman facilitated greatly improved access to intelligence and the construction of additional facilities, including a Sensitive Compartmented Information facility (SCIF). Working with the Israelis, he helped establish an F-21A Kfir Fighter Jet Aggressor Squadron at Yuma and the first "operator-to-operator" exchanges with the Israeli Air Force.

Kurth described his experience as an instructor and then his transition into ADT&E as follows:

The WTI course has an academic and then a flying phase. Despite featuring many first-class guest lecturers, the MAWTS-1 instructors made up the bulk of the academic instruction staff which progressed from generic topics to community and finally individual type/model/series specific topics over a couple weeks before flying.

MAWTS-1

Flying then progressed in the reverse order; specific, community and then generic (meaning full scale integrated exercises with air and ground aggressors).

Michael M. Kurth, Col, USMC (Ret.)
"Spot"

Col Kurth's direct involvement with MAWTS-1 gave him a unique perspective on its early leadership. He was a WTI student when Howard "Lopp" DeCastro was the first CO. He reported to MAWTS-1 as an instructor in 1981 when Bobby "Thunder" Butcher succeeded Lopp. He then served as an instructor for three years under Thunder and "Big Jake" Vermilyea. His final year at MAWTS-1 was spent in the standup of the Aviation Development, Tactics and Evaluation Department (ADT&E) with Randy "Dragon" Brinkley at the helm.

Access to computing power was very limited in the mid-1980s so instructors created their own media using photos of photos in magazines or professional publications annotated with hand created labels then converted to slides that were shown with a projector.

I presented two generic lectures during my tour; "Soviet Attack Helicopters" and "How to Make a Useful Presentation." My community lectures were on helicopter evasive and Air Comat Maneuvers (ACM).

The highest classification of these lectures was Secret, and this required my introduction to the Intelligence community, initially the Defense Intelligence Agency (DIA) and the Army's Foreign Science and Technology Center (FSTC) for subject matter.

I obtained a Top-Secret clearance and then, with the support of Secretary

Lehman, was one of several moving to ADT&E granted Sensitive Compartmented Information Clearances (SCI) where we were cleared to the highest level for threat Intelligence and technology development Special Access Programs (SAP). In effect, very operationally experienced officers were exposed to the best information on what both the good guys and the bad guys were doing.

This allowed the Marines at MAWTS-1 to get in front of the technology curve while working the training curve.

As Kurth put it:

This really did give us the time to do a little thinking about these new systems to prepare for their introduction. It also allowed us to provide some input based on things we had seen in operations and training.

We had a great feedback loop with the developers in industry and the labs because you know, none of this stuff got done without the labs and the test and evaluation units (VX-4 and VX-5) and the weapons centers across DoD.

So, we got a chance to talk about and think about how we would employ all these systems as they came online. And in some cases, we could offer improvements or modifications to what the developers were doing.

This was part of what enabled the so-called revolution in military affairs (RMA) almost ten years later because we were dealing with stealth, precision guidance and navigation, significantly improved weapons effects, unmanned systems, multi-spectral sensors and receivers, communications and networks, etc.

An element we discussed was working with foreign militaries to learn from their experience in combat.

One example was with the Israeli Defense Force (IDF) and lessons learned during the 1982 Peace for Galilee Operation. Here the Israelis crafted a wide range of innovations in how to attack Soviet equipment facing the Israelis, including how to improve the ability of attack helicopters to destroy armor, something clearly of interest to an attack helicopter pilot like Kurth.

Kurth noted:

I participated in the operator-to-operator exchange Secretary Lehman initiated. We spent a week in Israel visiting all the operating bases, looking at their weapons, looking at their capabilities and having discussions across the board, whether it was attack helicopter employment or fighter ops, deception operations, destruction of enemy air defenses (DEAD) or long-range strike.

It was an operator-to-operator relationship which opened the aperture on our thinking about innovation.

An interesting event occurred when we stopped in DC on our way back from Israel. "Dragon" had several names under consideration for his relief and he hadn't yet met one of them, so I let that guy know we were coming through town, and he offered to host the team at his house.

After a great evening "Dragon" pulled me aside and said "no need to look any farther" – the officer in question was Fred "The Assassin" McCorkle" who became the 5th MAWTS-1 CO.

Kurth's involvement in understanding foreign military systems was an important part of his tour.

As he noted:

While serving as an instructor at MAWTS-1, I became the DIA's Subject Matter Expert (SME) on Soviet helicopters and joined the European Tactics Analysis Team as the only rotary wing member analyzing Soviet tactics against helicopters.

Working closely with VX-5, who already demonstrated the feasibility and effectiveness of employing sidewinder missiles from the AH-1 Cobra, we developed air-to-air employment tactics.

We mounted a Tactical Aircraft Training System (TACTS) pod, essentially a dummy AIM-9, that could be tracked on the range emulating the weapon's performance in a "virtual" way just as the fighter community did for ACM training and debrief.

We then flew attack helicopters against both fixed and rotary wing aircraft for our edification and to provide informed advice to the US Army on their air-to-air field manual.

Another training innovation involved pilot training to prevail against electro-magnetic detection and guidance systems. Here the geographic location of MAWTS-1 was crucial.

The Marines could fly north into Navy and Air Force Electronic Warfare (EW) ranges to train. Training meant not just the employment of an aircraft's systems to survive but experiencing how their signature was observed from the actual systems used to kill them.

This allowed pilots to have a wholistic understanding of how to fight and win in an EW environment.

As Kurth underscored:

Pilots were gaining an understanding of what it was like for their opponents trying to kill them.

They were learning how to employ their countermeasures. How much and where the terrain can mask you – when and where you can you be seen. What does the radar horizon look like? And then for the anti-air defensive guns, what is the elevation range of the guns? This defines their lethal zone.

In short, MAWTS-1 training is rooted in understanding the evolving world of combat capabilities and how to shape standardized training to best positions the Marines for combat success.

And Kurth's own career and combat performance was a testimony to what MAWTS-1 contributed to the success of Marine Corps forces participating in Desert Storm.

Kurth's squadron, HMLA-369 with 18 AH-1Ws and 6 UH-1Ns, was the first Marine combat unit on the ground in Saudi Arabia in August 1990 assigned to defend against an expected Iraqi armored attack against coastal Saudi oil facilities. He had two MAWTS-1 instructors with him who provided exceptional desert and night operational training of the squadron's crews as well as the crews of follow-on squadrons.

Mike "Rifle" DeLong was the MAWTS-1 CO at the time. He basically mobilized the squadron to support the war effort in any way they could. While he was focused on maintaining the training mission in CONUS, he deployed any officer he could spare in support of USMC deployed operations – and they found themselves welcomed into the operational planning cells at each level of Marine command.

In an article published on 26 May 2016, Roger Showley highlighted Kurth's involvement in that war in an article entitled, "Pilot Flew 10 hours continuously in Persian Gulf battle."

Showley wrote:

Michael Kurth vividly remembers that day in February 1991. For 10 hours, he led squadrons of helicopters and fixed-wing aircraft through several miles of "thick black smoke that turned day into night," generated by Kuwaiti oil fields that Iraqi forces had set ablaze.

"I clearly remember the sound of burning oil wells when we got close to

them," he said. "You could hear the roar above the noise of the helicopter. The landscape seemed like something out of Dante's 'Inferno.'"

Kurth was credited with rallying his team to destroy as many as 70 Iraqi armored vehicles that day. Ten months later, President George H.W. Bush awarded Kurth the Navy Cross, one of two awarded during the Persian Gulf War. That recognition for combat heroism is second only to the Medal of Honor.

"I was just doing what I get paid to do, what the taxpayers pay me to do," Kurth said at the time. "I'm certainly gratified, but I'm also having a hard time putting it into perspective. I guess I just wish there was a better way to recognize everybody else."

The Navy Cross citation describes Kurth's bravery this way: "With total disregard for his own safety, he flew under and perilously close to high-voltage power lines. Placing himself at grave personal risk to intermittent Iraqi ground and anti-aircraft fire, Lt. Col. Kurth flew continuously for 10 hours during the most intense periods of combat, twice having to control-crash his aircraft."

Kurth said the fateful day in 1991 began at 5:30 a.m. and extended into the night. "There were a number of highlights, but what sticks with me the most is the teamwork, ingenuity, tenacity and sheer will of Marines," he said. "Nothing was going to slow any of us down as the environmental conditions were probably tougher than the enemy. I can only imagine what they — the enemies — thought as this very organized Marine Corps combined arms machine pursued them through this nightmarish smoke, oil, obstacles and minefields."

After the Persian Gulf War, Kurth earned a master's degree in national security and strategic studies in 1992 at the Naval War College and then chose to end his military service in 1996 at the rank of colonel.

Kurth said the Marine Corps and especially his time with MAWTS-1 taught him leadership, discipline, strategic thinking, and focus — assets that prepared him well for his work in the private sector after leaving military service. His advice to today's youths: "Always challenge yourself, be truthful and remember that people are relying on you to do the right thing."

MAWTS-1 and Trajectory Vision: Talking with the 5th Commander of MAWTS-1

December 13, 2023

We all know what tunnel vision is. But what is necessary for successful adaptation of a military in a dynamic situation is to have trajectory vision.

In other words, the ability to adapt but to do so with an eye on realistic adaptation which can be driven by new systems or new con-ops.

We had the chance to talk with LtGen Fred McCorkle about his involvement and experience at MAWTS-1 and how that shaped trajectory vision for the USMC which has been clearly demonstrated in adopting the Osprey and the F-35B and changing their con-ops to reflect doing so.

It is not just about adding some new technology; it is about understanding how best to leverage it to get the desired combat effect against a reactive enemy.

In our discussion with him, he underscored how the MAWTS approach was established and evolved to shape realistic innovation but with an eye to the future. And those who have been involved in the WTI experience were able to embrace change but had a "pretend I am from Missouri" kind of experience: show me.

This has led to MAWTS-1 being what we have referred to as an incubator for change. Or another way of putting it, rather than being trained to exercise tunnel vision, Marines have been trained to have trajectory vision or warriors who can embrace change. But not briefing chart change; real change demonstrated through what Lt General Rudder has called the "physics of warfare."

According to McCorkle: *I was a young major when first introduced to MAWTS-1. I went to the command, the DCA at the time was LtGen Fitch. When he first came, he said he was going to shut down MAWTS. But by the time he left his post, he had become one of the biggest supporters of MAWTS.*

We asked him, what did he think changed the DCA's mind?

McCorckle told us that the fact that MAWTS trained for integrated MAGTF operations was really the key.

MAWTS-1

He argued: *Those who came to the WTIs witnessed the best MAGTF training being done in the USMC.*

He credited John Lehman, then Secretary of the Navy, and the early commanders of MAWTs for setting in motion the framework to be able to provide for such integrated training. He also noted that Lehman allowed senior MAWTS officers, including himself, to receive clearances to have access to black Navy programs. This allowed them to have the possibility of trajectory rather than tunnel vision as commanders.

At the time, McCorckle noted that even the DCA's were not read into programs that the Admirals were allowed to be read into. This clearly created a problem as the introduction of new air systems and capabilities is a driving force for change.

One narrative McCorckle relayed was concerning the importance of the coming of night vision goggles. Although the night vision googles of the day were not perfect, McCorckle pointed out to general officers when he would have them experience low flights at night the following: *Would you rather have 20:50 vision with the NVGs or fly with 20:400 vision. They got the point.*

We then asked him about the safety of flight challenge.

McCorckle told his about his approach.

As CO of MAWTS I would brief every class and put up on the board a drawing. Here is this box. We want you to operate at the edge but not to go over the edge. You can operate near the corners of the box. I'll go to the mat for you to defend you if you operate that way. But if you go ¼ of inch outside of that box, I will be the first to "Recommend Pulling Your Wings"!

McCorkle also underscored the significance of MAWTS in terms of its COs having generally gone on to take General Officer Command and to bring their MAWTS experience forward and then re-invigorating MAWTS by contributing from their command position as well.

We then asked him about his Osprey experience and how MAWTS had prepared him for it.

He started by recounting a meeting with LtGen Fitch.

One day Fitch came up to me and put his arm around me and congratulated me on flying 65 different aircraft. I thanked him and told I was very humbled by the experience. He then underscored that he had flown 132 different aircraft!

Well, the Osprey was an aircraft like no other. Something which McCorckle experienced on his first piloting of the aircraft. It was certainly not a helicopter, a problem which still affects helicopter pilots who come to the Osprey and do not unlearn their rotorcraft skills.

As LtGen Fred McCorkle underscored: *At the end of a runway, a CH-46 will top end at 140 knots. If you are in loaded down Cobra, you are lucky to get to 120 knots at the end of the runway. And when I first flew it, I was at 250 knots at the end of the runway taking off.*

The test individual with me in the plane said we are limited in the speed we can go and you are already over the approved limit. I said the aircraft is virtually in neutral and we can go a lot faster and went to 330 knots.

This is what one can call trajectory vision at work

The 8th Commander of MAWTS-1: LtGen Barry Knutson Jr.

December 19, 2023

Barry "Knute" Knutson Jr. entered military service after graduating from the University of New Mexico in 1969. In 1986 he became the first commanding officer of Marine Fighter Training Squadron 401 and commanded it through its first two years as it introduced the Israeli built KFIR aircraft and stood up as the Marine Corps' first and only adversary squadron. He retired as a U.S. Marine Corps Lieutenant General.

He served as the eighth CO of MAWTS-1 from May 28, 1992, through June 3, 1994. During our discussion, he provided us with a wide-ranging perspective on the evolution of Marine Corps during his time in service. And the discussion of MAWTS-1 in that context really underscored a critical point – MAWTS-1 goes where it needs to work with whomever is solving problems which are critical to the USMC.

MAWTS-1

LtGen Barry Knutson Jr.

"Dragon" Brinkley had made the point to us that the Marines are the ultimate scavengers, who will go where they need to find what they need to be effective.

Throughout the interview, Knutson highlighted the Marine Corps focus which he was significantly involved in in training to operate against Soviet integrated air defense systems and to be able to do in order to provide the kind of ground support which was necessary.

During his time in MAWTS, they worked with the Tucson Air National Guard who were the specialists in low altitude flying. As Knutson noted: "Working with them enhanced our capability to do these missions significantly."

A major focus of attention was as well on EW in the context of being able to provide close air support. Knutson underscored that their focus on this core competence was worked with the USAF at Nellis and using the range at China Lake. Again, this is an example of MAWTS going where it was best to go to become proficient in a core competence.

A key development associated with MAWTS-1 is getting an

adversary squadron, namely the USMC fighter training squadron 401.

In a story written by Rick Llinares and published by *Naval Aviation News* in January-February 2002, the squadron was described this way:

On any day of the week, Marine Corps aircrews face off against some of the finest adversary fighter pilots in the world. They fight an "enemy" who probably has more experience and skill and who rarely makes mistakes.

Fortunately for them, they are fighting the Snipers of Marine Fighter Training Squadron (VMFT) 401. These experienced air combat tacticians are masters in the art of air combat tactics training.

Their job is to play the bad guy in order to teach Marine pilots about the threats they are likely to face in the world of aerial warfare. VMFT-401 is a reserve unit under the command of the 4th Marine Air Wing, based at Marine Corps Air Station (MCAS) Yuma, Ariz.

Activated in March 1986, it is the only dedicated adversary tactics training squadron in the Marine Corps, providing dissimilar air combat tactics instruction to both active and reserve Fleet Marine Force and fleet squadrons.

By June 1987 the first VMFT-401 Israeli-built F-21A KFIR fighters on loan from Israeli Aircraft Industries had been delivered, and the Snipers logged more than 4,000 accident-free sorties in support of 16 major exercises during the first year. In 1989, the squadron began transitioning to the Northrop F-5E Tiger II, an aircraft used by Air Force aggressor squadrons.[1]

We learned in our interview with Brinkley that Secretary Lehman played a key role in getting MAWTS the KFIR which allowed them to set up the aggressor squadron. Knutson who was at the meeting when Brinkley asked for Lehman's help provided his own recollection on the event.

Lehman came to dinner at MAWTS, and I attended as a LtCol. Lehman was very upset by the loss of Navy aircraft over Lebanon. He told Brinkley that MAWTS had done a lot to advance tactics for Navy aviation, but what do you need going forward? Brinkley said that he needed supersonic adversaries. And the very next day he created a deal with the Israelis to get 26 KFIRS shipped to the U.S. to play that role. I was then asked to stand up VMFT-401 using the KFIRS.

F-21A Kfir at Marine Corps Air Station, Yuma. Credit: Wikipedia

Knutson had a good offer to go to another squadron but found the new adversary squadron too interesting to turn down. The new squadron would have an active-duty Colonel in charge of the wing with 8-10 full time reservists and 8-10 weekend warriors. The maintenance department was supported by Israelis who moved to the area headed by an IDF colonel.

We concluded by focusing on the importance of the location in Yuma for MAWTS-1. They had their own ranges, including shared access with Luke AFB on the Goldwater range. They could fly North and work with USAF and U.S. Navy on their ranges and they could fly West and work with the West Coast Marines at Pendleton and use San Clemente Island.

In short, under Knutson's leadership and his fellow MAWTS CO's, MAWTS-1 contributed significantly to learning how to provide air support to Marine MAGTF's while confronting Soviet and Soviet like integrated air defense systems.

The Perspective of the 9th Commander of MAWTS: LtGen Castellaw

December 8, 2023

Laird had first met LtGen Castellaw when he was DCA. Laird at the time was working with for Secretary Wynne in defense acquisition, and met Castellaw in the context of the acquisition of the Osprey and the upcoming acquisition of the F-35B, the twin pillars of what we would call creating the Marines as three-dimensional warriors.

We had the chance to talk with him on November 17, 2023, to get his perspective on MAWTS and its role in the evolution of the Marine Corps.

LtGen Castellaw put it succinctly: *From my own experience, leading Marine and joint forces around the world, we needed the flexibility to provide different sets of tactics against different kinds of popup crisis elsewhere in the world. MAWTS helped inculcate in us that kind of mental agility the Marine Corps excels in, And I think at MAWTS we have that ability to have honest assessment of what we were doing and how to do it better or to avoid certain ways of doing things. Without that kind of brutal honesty, we will not succeed.*

As he thought back to students who came through MAWTS when he was there, they were all leaders who played a key role for the evolution of the Marine Corps air-ground team, such as LtGen Rudder, LtGen Davis and LtGen Heckl.

And one of the impressive things about MAWTS is that the former commanders are kept in the loop about the evolution of WTIs in the current courses. This provides for a feedback loop for excellence which is rather unusual in the world of combat.

According to Castellaw: *Going through MAWTS created a brotherhood in the Marine Corps. And that brotherhood was able to communicate from the same frame of reference throughout our careers, and several went on to take senior command positions. If you went through MAWTS, you learned what it took to make an aviation outfit and by extension the MAGTF truly capable.*

When asked what his major contribution was while being CO of MAWTS, he turned to their work on training and tactics for night vision operation of helicopters and fixed wing aircraft, a capability which would be so decisive in aviation operations in the Post-Desert Storm, Global War on Terrorism operations in Afghanistan and Iraq.

The Perspective of the 13th Commander of MAWTS-1: MajGen Raymond Fox

December 17, 2023

MajGen Raymond Fox had a distinguished career of more than 37 years in the United States Marine Corps, where he completed his military service as the Commanding General, II Marine Expeditionary Force and Commander Marine Forces Africa.

During his service career, Fox has held both staff and leadership roles, including combat operations during Operation Desert Storm, Operation Iraqi Freedom and Operation Enduring Freedom. In his last command, he was responsible for the welfare, training and logistics support for more than 58,000 Marines and Sailors, deployed around the world.

Fox has logged over 5,000 hours in Marine and Navy aircraft since receiving his commission through the Platoon Leaders Class and receiving his Naval Aviator designation. He is a graduate of the Australian Army Command and Staff College and the U.S. Army War College, and earned a master's degree in strategic studies from the U.S. Army War College, a Master's in public administration from Shippensburg University (Pennsylvania) and a Bachelor of Arts in political science from Eastern Washington University.

We had a chance to talk with Fox on December 4, 2023, and to get his perspective on MAWTS-1. He started by noting that he was the first CO of MAWTS to have never previously been a WTI instructor, (the first CO's who did not have time to be an instructor pilot) although he had gone through as a student.

A photo of the MAWTS 1 team in Kuwait hangs on the wall of the officer's bar at MAWTS-1.

He was CO from 1 November 2002 through 19 November 2004. This would fall in the period of Operation Iraqi Freedom and his Instructor Team would go to the war with 3D MAW, come back and do a WTI, 3 months after returning.

Fox went to MAWTS in the summer of 2002 to get ready to take over the command. But it was becoming clear that planes/forces were being sent to the Middle East in late 2022/2003 military operation so there was little point of having WTI instructors with no airplanes or students. According to Fox this meant most of the Instructor Pilots, C2 and Aviation Ground support Instructors Officer and Enlisted went to the Middle East. The next WTI was war. Fox went as a battle captain for General Amos and the 3D MAW.

Before they returned, Fox had the squadron members work on their lessons learned from the operation which they would take back as a basis for the WTI they would now do in the Fall of 2003.

MajGen Fox noted: *This would be essential in this WTI for our assessment of what weapons and tactics were working and which were not working. We had a wealth of operational knowledge to work from.*

They briefed their findings to the Air Force at Nellis and the

Navy at Fallon. Fox noted: *The Air Force was receptive; the Navy was less receptive.*

He emphasized that in this period they worked integration of the TTP manuals with the Air Force and the Navy which is a key effort which three schools still pursue.

He noted that they set up the first Desert Talon exercise in January 2004. Coming from MEU background, Fox was used to workups for integrated operations on deployment. Desert Talon exercises were focused on integration of Marine Corps forces going to Iraq. This meant 2 WTI classes a year, 2 Desert Talon's a year and Constant trips to Iraq to stay up to date on threat and TTPS.

U.S. Marine Corps Maj. Gen. Raymond C. Fox, commanding general of II Marine Expeditionary Force, salutes during his retirement ceremony at Camp Lejeune, N.C., July 16, 2014. Fox retired after 37 years of honorable and faithful service to the corps. U.S. Marine Corps photo by Sgt. Gabriela Garcia/Released

Fox mentioned that the City of Yuma, and Yuma County leadership let them do convey protection exercises whereby Marine Air ran air cover for conveys and did so in Yuma itself. As Fox noted: *This allowed the ground side and the air side to learn how to work together for convoy protection.*

As he looked back at his time at MAWTS-1 he emphasized that: *The greatest thing about MAWTS is the people. You generally get the best the USMC has to offer in each area of the Marine Corps. And generally, we got great support from the joint world as well. The previous MAWTS-1 Commanders, deliver mentoring and guidance.*

He also mentioned the role which Post Command rotary/fixed

wing colonels played as providing an "extra set of eyes and ears to help out" During the flight phase, this idea was generated by Jon Davis who would replace him.

MajGen Fox concluded with a key point.

The CO of MAWTS controls, the red, the blue and the white – the whole battlefield. This allows for very innovative scenarios and training for the students. And the WTI instructors and the students in the FINEX get to experience first-hand what combat integration and adversary efforts to break up an integrated force is all about.

Turning Training into New Combat Capabilities

December 12, 2023

We did a follow up interview with Randy "Dragon" Brinkley on November 17, 2023, to focus on how MAWTS training turned into new combat capabilities which were significant game changers.

The first example was working with Night Vision Googles with the rotorcraft force which led to success in the Gulf War as they Marines successfully prosecuted Iraqi forces using this combat advantage capability. Without introducing them at MAWTS and training with the evolving NVG capabilities, the Marines would have to do it the Vietnam War way, seat of the pants combat learning.

"Dragon" underscored that with the failure of the desert one operation in Iran, there was real concern to avoid another high-risk operation like that one. This meant that there was pressure to attenuate rotorcraft operational experimentation and preparation for another force insertion operation of the sort that would happen more than a decade later in Iraq.

But when Col. Jake Vermilyea, a helicopter pilot, came to MAWTS, first as the XO to Col Butcher and then as the CO of MAWTS-1 (1982-1984), he pushed the envelope on nap of the earth training using NVGs.

The DCA at the time was LtGen William Fitch was who did not support at MAWTS at the outset of his time as DCA but during his time he participated in WTIs and saw the key role which MAWTS was playing for MAGTF integration.

In fact, if you look at Fitch's oral history you can see his attitude when he was CO 1st MAW.

Yes, MAG-36 was a helicopter group. One of the weak links that I found was the WTIs, the Weapons Tactics Instructors. In those days they had an air of arrogance about them. And while we had – this was the early days of WTI – we would have WTI instructors who kind of thought they were supermen.

But then later he clearly had changed his mind as reflected in this comment in his oral history.

MAWTU would go out of business after MAWTS came into being at Yuma. MAWTU kind of did its own thing, they got their students in and did the various and sundry kind of training that they needed to do. Of course, MAWTU had an officer in charge, and it was his job to train those students that were assigned to his school.

But in the case of MAWTS, that was probably one of the smartest moves Marine aviation ever did and it's been the picture book testimonial to having that kind of a structure for training.

And I was talking to John Cox the other day, who was the former CG of 3rd Wing, and he was talking about – in fact it was at the Commandant's reception here last Friday night – he was commenting about how many MAWTS commanders had made brigadier general and there are only two or three that haven't made brigadier out of that job.[2]

"Dragon" commented: *His support was a key turning point in USMC emphasis on training for nap of the earth and NVG training. The training was then included in the WTI syllabus and training was now pursued in the standardized way in which we have pursued air combat training and across the Fleet Marine Force.*

The second example he gave was of training for low altitude tactics. He made the point that the Marines are the ultimate scavengers and look wherever they can to gain a combat edge. In this case, it was MAWTS looking to the Tucson Air National Guard.

He noted: *I went over to the 174 Tactical Fighter Training Group in Tucson which consisted of F-100 pilots who had flown during the Vietnam War and had extensive experience with low altitude tactics.*

They focused on attacking an integrated air defense system by going low on the egress but fly high when in the target area where you would deliver your weapons and avoid small arms fire in the target area. They had 1000s of hours of flying at 100 feet and we learned a lot from them, and which led to success in the Gulf War.

The third example was working with foreign militaries to learn tactics which advanced how the Marines could be more effective and successfully. The example he provided was of Secretary Lehman setting up an exchange program with the IAF.

As "Dragon" commented: *The Israelis then sent a fixed wing and rotary wing student to a WTI course, which we changed the syllabus to reflect the proper security level. The IAF A-4/F-15 pilot shared with us the planning and tactics employed in attacking a target protected by an integrated air defense system and the rationale for those tactics.*

Among other things the IAF attendees also shared with us how they used their UAVs to push video into their Cobras and F-15s. When Lehman decided that the Navy would procure Pioneer UAVs, the MAWTs experience was important.

The IAF WTI attendees also reinforced from their combat experience what we were learning from the 174th, namely fly in low on egress but fly high when delivering your weapons. We certainly subsequently benefitted from that knowledge and experience during Desert Storm combat operations.

He concluded: *By participating in MAWTS, Marine Corps officers learned how to work combat integration and employment in support of the MAGTF.*

You must have your C2, and your logistics planned and fully integrated to be successful and you exercise that in the WTI FINEXs. Learning in an integrated environment and the warfighters level, that is the focus and overall objective of MAWTS.

What MAWTS-1 Contributes to the Force

December 4, 2023

We had a chance on November 16, 2023, to talk with Lt General Steve "Stick" Rudder concerning how during his career MAWTS-1 contributed to the ongoing innovation and combat capabilities for the USMC. LtGen Rudder was the MARFORPAC commander in his last posting, but before that he was DCA.

LtGen Rudder identified a number of MAWTS-1 contributions. He noted that he came first to MAWTS-1 as a WTI instructor at MAWTS-1 after his involvement in Desert Storm and his generation of Marines brought that experience directly into the course.

He characterized the general MAWTS contribution as follows: *MAWTS-1 provides the only opportunity for future tactical and operational junior leaders to come together and expose them to how the Marine Corps fights as an integrated MAGTF and not as separate communities.*

As a young officer, you experience participating in the fully integrated MAGTF and the complexity and strength of the integrated force. You have the unique opportunity to learn how to be a tactical leader in the broader context.

Rudder talked about what he called the advantages of a "MAWTS product."

He characterized the product as follows: *A Marine WTI who can go back into squadrons and make them combat proficient as a unit and operationally relevant in a multitude of environments.*

The WTI provides leadership for that squadron to make sure it can safely execute the missions which they have been assigned. He or she is also there to groom future leaders for Marine Aviation.

Rudder provided a case in point of how MAWTS contributes to the innovation process. He noted when he was DCA, he focused significantly on digital interoperability, mesh networks and the general effort to integrate the Marines more effectively as a kill web force.

He noted: *The ideas from MAWTS started that process. Whether it was night vision goggles, MV-22 tactics, F-35 integration, new tactical tablets or MAGTABs, or testing new command and control technology, MAWTS led the way in technology, proof of concept, and follow on training and standardization.*

He underscored: *Unlike other schoolhouses, MAWTS is the only organization that can keep pace with technology and changes in the operational environment.*

Lt. Gen. Steven R. Rudder (left), commander, U.S. Marine Corps Forces, Pacific, meets with Gen. Izutsu Shunji, Chief of Staff, Japan Air Self-Defense Force, at the Japan Ministry of Defense, Tokyo, July 7, 2021. Photo by U.S. Air Force SSgt Samuel Burns.

But how does it do so? It is not through war games or briefing slides – it is through actual testing in training what Rudder referred to as the "physics of combat." Through a direct connection to the fleet Marines or the warfighting lab, when a new tactical concept is introduced, MAWTS tests how such a concept would be executed in the real world.

How does the physics of force sustainment, force maneuver, force support, force protection and delivering of effects work or not?

The challenge in Rudder's words: *We need to avoid smoke and mirrors. The talking points from a wargame or ideas from force design planners may sound good, but it has to have an operational look. An A plus for your ideas may equal a D for practical execution. If the concept or design does not make it through the physics of operational execution, the idea needs to be rejected or seriously modified. Giving problems and operational concepts to MAWTS is crucial to working the physics of combat in testing out new tactical approaches.*

A Conversation With the 14th Commander of MAWTS-1: LtGen "Dog" Davis

January 27, 2024

In January 2024 we had a chance to do our final interview with the former commanders of MAWTS-1, namely with LtGen "Dog"-Davis. He was the CO of MAWTS-1 from 19 November 2004-7

July 2006 and had served as the XO for Col later MajGen Raymond Fox, his immediate predecessor.

Fox had become the CO of MAWTS-1 just prior to Operation Iraqi Freedom in 2003 and he along with most of MAWTS-1 went to that operation. After the war, they set up the first Desert Talon exercise which was focused on the integration of the Marine Corps force going to Iraq.

MajGen Fox underscored the key role which MAWTS-1 plays in training the trainers and driving innovation for an integrated USMC.

The CO of MAWTS controls, the red, the blue and the white – the whole battlefield. This allows for very innovative scenarios and training for the students. And the WTI instructors and the students in the FINEX get to experience first-hand what combat integration and adversary efforts to break up an integrated force is all about.

LtGen Davis built upon this perspective and underscored how MAWTS-1 was pushing the envelope of innovation as a core function of its approach to training. And because through the WTIs they were training the squadron trainers they were diffusing the drive for innovation forward.

Of course, Davis brought his own background and experience to the command. He was a Harrier pilot and had served with the Brits in West Germany and had learned their way of operating Harriers in a high-end Cold War threat environment every day. He also had written his master's thesis in 1994 on the challenges of operating in an urban environment which became very relevant in the years to come in Iraq.

In his discussion with us, a major theme was how to work effective ways to weaponize the force. Force integration was key, but which weapons, for which missions and how to make sure that the force was effective as an integrated force capable of delivering the desired effects. The approach here was to provide flexibility to the force but to ensure that the force worked to deliver the desired effects which is a way of describing effective weaponization of the force.

Another key theme was operational effectiveness and operational excellence. On the effectiveness side, the transition to a predominance of precision guided munitions (PGMs) for the MAGTF and ACE and meant they adjusted the WTI course focus to emphasize the use of and optimization of that ordnance for throughout every sortie in the course.

On the operational excellence side of the ledger, Davis was tasked to operationalize safety at WTI and to reduce the mishap rate associated with WTIs in the fleet. Davis asked for an additional three days to be added to the WTI course academics and built a Tactical Risk Mitigation syllabus that helped WTIs understand risk and develop strategies to help their units avoid the "Blue" Threat and focus more on mitigating the Red Threat to more effectively support the MAGTF.

While he was at MAWTS-1, they were anticipating the arrival of the Osprey and after that the F-35. This has been a virtual revolution in USMC operations and thinking. But "Dog" felt strongly that the Marines needed fifth gen to operate in the kind of environments which were clearly emerging with higher-end competitors.

LtGen "Dog" Davis during our 2017 interview in his office in Cherry Point when he was the head of Second Marine Air Wing.

They had done exercises which demonstrated that against a high end nation versus nation threat (not counter insurgency) only using 4th gen and lower legacy aircraft, the Marines would lose significant numbers of aircraft and not have the desired results against the target (mission fail).

As Davis noted:

In one strike we put 34 strikers against an integrated air defense system (air and ground threats) and we lost 1/2 the FA-18, Harrier and Prowler strike force and no one hit the target. Mission Fail. Therefore, I asked the USAF commander at Nellis to lend the WTI strike package 6-8 F-22s to run the same scenario next course - to expose the WTI students and MAWTS-1 staff to 5th generation capabilities and to understand how they could enhance the survival and lethality of the MAGTF in the near future.

The next class's strike had identical strikers and scenario with the addition of 8 F-22s. We killed all the bad guy fighters, destroyed or degraded the SAMS, achieved 100% mission success 8n the target and lost zero aircraft. It was an epiphany for the MAWTS-1 staff. After working with the F-22, it was obvious that fifth gen capabilities were necessary for the Marines to win in the nation state versus nation state threat environment.

Lastly, Davis revamped the MAWTS-1 Air Officer Course to focus it on building MEU and Regimental Air Officers with the knowledge and skills needed to ensure that those units received the very best from their Air Officers, ultimately leading to a new MOS for MAWTS-1 certified Air Officers and making a pre-requisite to hold that billet.

The focus of MAWTS-1 on driving a way ahead for innovation for the Marines operating as an integrated force but incorporating new systems, new capabilities, standards and new con-ops was underscored by LtGen Davis as a key element in generating the kind of MAGTF needed to win the battles that loomed on the nation's bow.

10

LtCol DeCastro Sums It Up

December 13, 2023

Howard DeCastro, LtCol USMC, (Ret.)

I was invited by Colonel Kelvin Gallman to be the guest of honor at the MAWTS-1 Marine Corps Ball in 2019. An edited version of the speech which I gave follows:

During the Vietnam conflict, Marine Air conducted effective deep strikes on enemy targets, and transport and close air support for our Marines on the ground. However, we made a lot of mistakes and our use of airpower was well short of optimum.

We didn't have a consistent strategy, we lacked up-to-date targeting intelligence, we rarely took full advantage of the capabilities that come with skillful integration of forces, and our ability to effectively support our Marines on the ground was often hampered by poor communication, by the Ground Commander's lack of knowledge of our capability, and sometimes even by a lack of trust.

The result was, more Marines died in the air and on the battlefield than would have if we had done a better job.

Project 19 highlighted some of the problems and provided the opportunity to improve. With lots of thought and hard work by a

MAWTS-1

number of Marines at Headquarters and in the Fleet Marine Force, MAWTS -1 was born more than 41 years ago.

The goal was to improve the effectiveness of Marine Air and provide the best possible support to our Marines on the ground. To do that we knew it was important that every aviator have a clear understanding of their own capability and the capabilities of other aviation elements, and to learn how to best integrate those capabilities to support our combat troops.

In June 1978 we started with 35 MAWTS Instructors and a support staff of 15 Marines. You have increased in size, now up to 249 staff, one short of five times our original number.

More importantly, you have grown dramatically in capability to become the finest aviation training and tactics development organization in the world. There is no doubt that you are exceeding the expectations that were set when MAWTS was established.

The addition of 401 provided another vital and dramatic improvement.

After General Amos retired as Commandant, he and I had a chance to talk. He told me that MAWTS literally saved Marine Corps Aviation. I suspect that was an exaggeration, but, without question, MAWTS has dramatically improved Marine Aviation and our working relationship with the Ground Forces.

And that means fewer Marines will die on the battle field.

The challenge for us 41 years ago, and the challenge for you today, is to continually improve. To improve takes an understanding of enemy capabilities and projecting their future capabilities so we can develop the hardware, software, tactics, and our own capabilities that keep us always in the lead.

Every one of you should be proud to have earned the right to be assigned to MAWTS, and you should be proud of what you are accomplishing.

I was going to give you a motivational speech, but then I realized you are just like the original members of MAWTS-1.

You are completely self-motivated, are never satisfied with the status quo, and you will keep thinking, keep communicating, keep innovating, and keep pushing each other to make Marine Air and the Marine Air Ground Team better every day. That is who you are, and you know how important it is to keep getting better.

You know that the Marines on the ground trust you and are counting on you.

For everything you have done and all you do, I thank you!

Happy Birthday Marines - Semper Fidelis

Appendix: The Commanders of MAWTS-1

LtCol Howard L DeCastro
 1 June 1978-27 June 1980

Col Bobby G, Butcher
27 June 1980-5 August 1982

Col. Clyde L. Vermilyea
5 August 1982-8 June 1984

MAWTS-1

Col Randolph H. Brinkley
8 June 1984-11 July 1986

Col Fred McCorckle
11 July 1986-9 June 1988

Col Michael D Ryan
9 June 1988-15 June 1990

Col Michael P. Delong
15 June 1990-28 May 1992

MAWTS-1

Col Bruce Barry Knutson Jr.
28 May 1992-3 June 1994

Col John G, Castellaw
3 June 1994-17 May 1996

Col Keith J. Stalder
17 May 1996-15 May 1998

Col William D. Catto
15 May 1998-10 May 2000

Col Martin Post
10 May 2000-1 November 2002

Col Raymond C. Fox
1 November 2002-19 November 2004

Col Jon M. Davis
19 November 2004-7 July 2006

Col Robert F. Hedelund
7 July 2006-27 June 2008

MAWTS-1

Col Gary L. Thomas
27 Jun 2008-18 June 2010

Col Karsten S. Heckl
18 June 2010-8 December 2011

Col Bradford J. Gering
8 December 2011-22 May 2014

Col James H. Adams
22 May 2014- 12 May 2016

Col James B. Wellons
12 May 2016-21 May 2018

Col Kelvin W. Gallman
21 May 2018-8 November 2019

Col Steve E. Gillette
8 November 2019-12 May 2022

Col Eric D. Purcell
12 May 2002-3 May 2024

The Commanding officer of MAWTS-1 when we wrote the book. Col Purcell showed us the support and leadership which helped us finish our effort. For which we thank him.

Col Joshua Smith 3 May 2024

The outgoing and incoming CO of MAWTS-1 at the Change of Command ceremony, May 3, 2024. Credit: Robbin Laird

The past and current Commanding Officers of MAWTS-1 attending the Change of Command Ceremony, May 3, 2024. Credit: MAWTS-1

About the Authors

Dr. Robbin F. Laird

A long-time analyst of global defense issues, Dr. Laird has worked in the U.S. government and several think tanks, including the Center for Naval Analysis and the Institute for Defense Analysis.

He has written several books dealing with the USMC including, *The U.S. Marine Corps Transformation Path: Preparing for the High-End Fight (2022)*.

He is a Columbia University alumnus, where he taught and worked for several years at the Research Institute of International Change, a think tank founded by Dr. Brzezinski.

He is a frequent op-ed contributor to the defense press and has written several books on international security issues.

Dr. Laird has taught at Columbia University, Queens College, and Johns Hopkins University.

He has received various academic research grants as well from various foundations, including the Thyssen Foundation, the National Science Foundation. and the United States Institute for Peace.

Dr. Laird has worked for many elements of the U.S. government and with think tanks such as The Center for Defense Analysis and the Institute for Defense Analysis.

He is a member of the Board of Contributors of *Breaking Defense* and publishes there on a regular basis.

He is a frequent visitor to Australia where he is a Research Fellow with The Richard Williams Foundation in supporting their seminars on the transformation of the Australian Defence Force. He has published three books on Australian defense issues.

He is also based in Paris, France where he regularly travels throughout Europe and conducts interviews and talks with leading policy makers in the region.

Edward Timperlake

Ed Timperlake began his Marine flying career after graduating TBS (The Basic School) in December 1969. Very early in his training he ejected out of an exploding jet on a night solo on takeoff below 1000 feet. Surviving that event, he made a commitment to continue to fly as long as the Marines would let him.

Seventeen years later he finished his career as Commanding Officer VMFA -321 and he accumulated slightly over 3000 hours, of which 2000 were in the F-4 Phantom and the remainder in A4-E Skyhawk

When medically recovering from being medically evacuated from "The Rose Garden," he was allowed to fly co-pilot in the UH-1E for 200 hours where he discovered the fun of flying helicopters.

Other Second Line of Defense Books of Interest

All of these *Second Line of Defense* books can be purchased in e-book or paperback versions on Amazon or other on-line booksellers.

A Maritime Kill Web Force in the Making: Deterrence and Warfighting in the 21st Century

By Robbin Laird and Ed Timperlake

As Vice Admiral (Retired) Dewolfe Miller underscores: "Ultimately, peer threats are what drives change and inspires clarity in the way the Navy mans, trains, and equips its forces to defend freedom and deter aggression on a global scale. "A Maritime Kill Web Force in the Making" is the story of the evolution of the US Navy and its preparation for high-end warfare.

This is essential because the future of combat is to bring trusted and verifiable assets to the fight. The emphasis has been on connectivity, accelerated tactical decision making, as well as common equipment, that allows integration of systems within single services, across services and into allied services in a deliberate and disciplined manner. This publication provides a timely reminder of why the transformation of today's force is so necessary."

Other Second Line of Defense Books of Interest

Published in 2022.

The U.S. Marine Corps Transformation Path: Preparing for the High-End Fight

By Robbin Laird

The United States Marine Corps began its modern transformation path after the introduction of the Osprey in 2007. In a series of in-depth interviews with the United States Marines, this analysis highlights the transformation strategy that has made the USMC one of the most dynamic military forces in the world today.

From the land wars to dealing with peer competitor threats and engagements, this book demonstrates how the Marines are navigating the strategic shift to craft innovative solutions for the return of Great Power competition.

Many coalition partners look to the USMC as a relevant benchmark for the kind of multi-domain operations which they can pursue.

For many allies, their force structure approximates the size of the USMC, and they find the fit better than emulating the total force which the United States has built.

It is also the case that the legacy force coming out of the land wars is not directly applicable in terms of its warfighting relevance to the approaches for combat with the peer competitors.

"Only time will tell how the Marine Corps navigates this treacherous transformation journey, but it's not the equipment that will make the Corps successful on the future battlefield – it's the Marines."- LtGen George Trautman, USMC (Ret).

Published in 2022.

The Role of the Osprey in the Pivot to the Pacific

By Robbin Laird

The Osprey provides an important stimulant for the shift in con-ops whereby the Navy's experimentation with distributed operations intersects with the U.S. Air Force's approach to agile combat

Other Second Line of Defense Books of Interest

employment and the Marine Corps' renewed interest in Expeditionary Advanced Base Operations (EABO).

In other words, the reshaping of joint and coalition maritime combat operations is underway which focuses upon distributed task forces capable of delivering enhanced lethality and survivability.

The U. S. Navy's deployed fleet — seen as the mobile sea bases they are – faces a significantly different future as part of a distributed joint force capable of shaping a congruent strike capability for enhanced lethality.

This means not only does the fleet need to operate differently in terms of its own distributed operations, but also as part of modular task forces that include air and ground elements in providing for the offensive-defensive enterprise which can hold adversaries at risk and prevail in conflict.

But how did we get here in 2023? How has the strategic shift for the joint forces evolved and caught up with what the tiltrotor revolution has enabled? And how has the Osprey evolved since the recognition of great power competition by the Trump Administration in 2018?

It began as a pivot to the Pacific in 2013; it is becoming a con-ops revolution enable in part by tiltrotor aircraft. The book takes two snapshots of this transition.

The first focuses on the introduction of the Osprey into the Pacific when the Obama Administration announced its "Pivot to the Pacific.

The second focuses on changes to the tiltrotor enterprise since 2019 after the Trump Administration highlighted the "Great Power" competition.

Published in 2023.

The Coming of the CH-53K: A New Capability for the Distributed Force

By Robbin Laird

This book describes the coming of the CH-53K Kilo to the USMC and to its first international customer, the Israeli Defence

Other Second Line of Defense Books of Interest

Force. It is based on extensive interviews with the persons involved in the development, testing, build, and maintenance of the new combat air system.

For air system it is -- built by the digital thread development and manufacturing approach, the aircraft is designed with maintainability and fleet support in operations as a key focus of the program,

If it were called CH-55 instead of the CH-53K perhaps one would get the point that these are very different air platforms, with very different capabilities.

What they have in common, by deliberate design, is a similar logistical footprint, so that they could operate similarly off of amphibious ships or other ships in the fleet for that matter.

But the CH-53 is a mechanical aircraft, which most assuredly the CH-55 (aka as the CH-53K) is not.

In blunt terms, the CH-55 (aka as the CH-53K) is faster, carries more kit, can distribute its load to multiple locations without landing, is built as a digital aircraft from the ground up and can leverage its digital backbone for significant advancements in how it is maintained, how it operates in a task force, how it can be updated, and how it could work with unmanned systems or remotes.

These capabilities taken together create a very different lift platform than is the legacy CH-53E. In a strategic environment where force mobility is informing capabilities across the combat spectrum, it is hard to understate the value of a lift platform, notably one which can talk and operate digitally, in carving out new tactical capabilities with strategic impacts.

The lift side of the equation within a variety of environments can be stated succinctly. The King Stallion will lift 27,000 lbs. external payload, deliver it 110 nm to a high-hot zone, loiter, and return to the ship with fuel to spare.

What that means is JLTV's (22,600-lb.), up-armored HMMWV, and other heavier tactical cargos go to shore by air, rather than by LCAC or other slower sea lift means. For less severe ambient conditions or shorter distances than this primary mission, the 53K can carry up to 36,000 lbs.

With ever increasing lift requirements and advancing threats in

Other Second Line of Defense Books of Interest

the battlefield, there is no other vertical lift aircraft available that meets emerging heavy lift needs.

There are a lot of platforms that can blow things up or kill people, but for heavy lift, the CH-53K is the only option. The digital piece is a foundational element and why it is probably better thought of as a CH-55. This starts with the fly-by-wire flight controls. The CH-53K is the first and only heavy lift fly-by-wire helicopter.

The CH-53K's fly-by-wire is a leap in technology from legacy mechanical flight control systems and keeps safety and survivability at the core of the Kilo's design while providing a portal to an optionally piloted capability and autonomy.

The CH-53K's fly-by-wire design drastically reduces pilot workload and minimizes exposure to threats or danger, particularly during complex missions or challenging aircraft maneuvers like low light level externals in a degraded visual environment allowing the pilot to manage and lead the mission vice focusing on physically controlling the aircraft.

What this means is that the CH-53K "can operate and fight on the digital battlefield."

And because the flight crew are enabled by the digital systems onboard, they can focus on the mission rather than focusing primarily on the mechanics of flying the aircraft.

This will be crucial as the Marines shift to using unmanned systems more broadly than they do now.

Published in 2023.

My Fifth-Generation Journey: 2004-2018

By Robbin Laird

Tis is the first of two books looking at the standup of the F-35 global enterprise.

According to the author: "It is my personal journey observing the development and evolution of the aircraft from my time working with the man who coined the term fifth-generation aircraft, Michael W. Wynne, through my many visits to F-35 sites, interviews with pilots, maintainers, and U.S. and allied government officials who navigated through the incredibly negative press and government officials trying to kill the program to have delivered a unique capability in the history of combat aircraft."

"It is a personal journey and I take the reader to many of the places where I went to talk with the F-35 nation. But I did so not only in the United States but in the nations of key members of the F-35 global enterprise. My journey is unique and I tell it not because of a desire to be remembered but in playing a role of recorder of history of those members of the military of many nations who made this capability real despite the media and many government officials desires to see them fail."

"But failure would have meant that we would have had even less capability than we have now after 20 years of fighting in wars of "stability" which have brought us the opposite."

As LtGen George Trautman, USMC (Ret), Former USMC Deputy Commandant for Aviation, writes in the forward to the book:

"Robbin's ability to capture the perspectives of the key players, from pilots to maintainers and logisticians, provides a comprehensive and insightful account of this revolutionary aircraft.

"Moreover, the book uncovers the geopolitical implications of Fifth-Generation warfighting capabilities. As nations seek to assert their dominance and secure their interests, the strategic implications of these technologies ripple across the global stage.

"Through a series of personal essays and skilled interviews with those who understand the aircraft, *My Fifth-Generation Journey: 2004-2018* deftly navigates through this complex web, painting a vivid picture of how Fifth-Generation warfighting will shape the future geopolitical landscape."

The book includes a number of original photos shot by the author during his visits highlighting the roles of pioneers in the setting up of the F-35 in the services and abroad.

The book was published in 2023.

Other Second Line of Defense Books of Interest

The Coming of Maritime Autonomous Systems: Empowering and Enhancing the Kill Web Force

This book addresses the coming of maritime autonomous systems to the U.S. Navy and allied fleets. They are obviously coming as well to adversarial forces as well.

It is clear that maritime autonomous systems will become a key part of the modernization of the United States (U.S.) and allied security and combat forces. They will also become part of the adversarial forces and will need to be countered as well.

Even though maritime autonomous systems are already here and ready to be deployed. there are the vested interests of maintaining a legacy approach to maritime operations and the thinking associated with such an approach which impede the way ahead.

This book lays out some key developments which have already occurred regarding maritime autonomous systems and what interactive changes between the arrival of such systems and the evolution of concepts of operations might be anticipated.

One such change is the emergence of a new class of ships, which might be called mother ships, and that dynamic is discussed later in the book.

Maritime autonomous systems are not ends unto themselves, but capabilities which enhance the distributed maritime force and its ability to contribute to joint or coalition operations across the spectrum of warfare.

Ranging from weapons to C2 to ISR to logistical payloads, maritime autonomous systems can deliver expanded capabilities to a Navy usually measured in terms of numbers and tonnage of capital ships.

In short, it is a whole new entry point into the future which empowers the force and provides for enhanced capabilities. But to do so is not just about technological development; it is about evolution of concepts of operations and evolving C2 and ISR working relationships for the fleet.

As LtGen (Retired) Steve Rudder, the former MARFORPAC Commander noted in the forward to the book:

Other Second Line of Defense Books of Interest

"Dr. Robbin Laird has been leading the reporting on Unmanned Systems and Kill Webs for many years and has been producing forward thinking pieces on the evolution of autonomy.

"At each achievement, whether it be Ukraine, TF-59 in the Arabian Gulf, or the Australian Defense Force, his articles and books have provided a window into the future dominance of autonomous maritime systems and the journey into the Kill Web."

"The reader of Robbin's book should walk away with a sense of how autonomous maritime systems are changing how we think about Naval Warfare."

The book was published in 2024.

Notes

Introduction

1. https://www.pprune.org/archive/index.php/t-206962.html.
2. Katie Lange, "It Started in a Parking Lot: TOPGUN's History Revealed," *DOD News* (January 26, 2022), https://www.defense.gov/News/Feature-Stories/Story/Article/2823132/it-started-in-a-parking-lot-topguns-history-revealed/
3. https://www.nellis.af.mil/About/Fact-Sheets/Display/Article/2605882/414th-combat-training-squadron-red-flag/
4. https://sldinfo.com/2016/03/squadron-fighter-pilots-the-unstoppable-force-of-innovation-for-5th-generation-enabled-concepts-of-operations/. Also see, my co-authored study, J.R. Hiller and E.T. Timperlake. "Exploratory Models of Pilot Performance in Air-to-Air Combat," *Naval War College Review* (Vol. 34, No. 1, January-February 1981), pp. 82-92.
5. Ed Timperlake, "Shaping a Way Ahead to Prepare for 21st Century Conflicts: Payload-Utility Capabilities and the Kill Web," *Second Line of Defense* (September 14, 2017), https://sldinfo.com/2017/09/shaping-a-way-ahead-to-prepare-for-21st-century-conflicts-payload-utility-capabilities-and-the-kill-web/
6. Ed Timperlake, "Shaping a Way Ahead to Prepare for 21st Century Conflicts: Payload-Utility Capabilities and the Kill Web," *Second Line of Defense* (September 14, 2017), https://sldinfo.com/2017/09/shaping-a-way-ahead-to-prepare-for-21st-century-conflicts-payload-utility-capabilities-and-the-kill-web/

1. The 2011 and 2012 Visits: The Coming of the F-35B

1. "The Future is Now": Clearing the Decks for the iPad Generation Pilots," *Second Line of Defense* (October 4, 2011), https://sldinfo.com/2011/10/the-future-is-now/
2. Robbin Laird, "An Update on the F-35 Integrated Training Center at Eglin AFB," Second Line of Defense (September 16, 2012), https://sldinfo.com/2012/09/an-update-on-the-f-35-integrated-training-center-at-eglin-afb/
3. Robbin Laird, "The F-35 and Pacific Strategy: Shaping a Core Lynchpin," *Second Line of Defense* (November 8, 2012), https://sldinfo.com/2012/11/the-f-35-and-pacific-strategy-shaping-a-core-lynchpin/
4. "Captain Hall Discusses the USS America: Looking Towards the Future," *Second Line of Defense* (November 3, 2012), https://sldinfo.com/2012/11/captain-hall-discusses-the-uss-america-looking-towards-the-future/
5. "The impact of the USS America on USMC Operations: "A MAGTF on Steroids," *Second Line of Defense* (November 21, 2012), https://sldinfo.com/2012/11/the-impact-of-the-uss-america-on-usmc-operations-a-magtf-ace-on-steroids-usmc-operations-a-magtf-on-steroids/

Notes

2. A 2013 Perspective: The Yuma Incubator of Change

1. Robbin Laird, Edward Timperlake, and Ricard Weitz, *Rebuilding American Military Power in the Pacific: A 21st-Century Strategy* (Praeger Security International) (p. 5). ABC-CLIO. Kindle Edition.

3. The 2014 Visit

1. What does named mission mean? See the following:
   https://www.govtech.com/em/disaster/how-pentagon-names-military-operations.html

4. The 2016 Visit

1. "VMFA-121 Departs for Relocation to Japan," *USMC* (January 10, 2017), https://www.miramar.marines.mil/News/Press-Release-View/Article/1046222/fy17001-vmfa-121-departs-for-relocation-to-japan/
2. https://sldinfo.com/2010/02/the-osprey-in-afghanistan-a-situation-report-2/.

5. The 2018 Visit

1. The source quoted is no longer located at the location used in the original article which was as follows: https://marinecorpsconceptsandprograms.com/programs/command-and-controlsituational-awareness-c2sa/common-aviation-command-and-control-system.
2. "Working the Integrated Fires Mission at Fort Sill," *Second Line of Defense (April 4, 2018)*, https://sldinfo.com/2018/04/working-the-integrated-fires-mission-at-fort-sill/
3. "The US Army, Innovation and Shaping a Way Ahead for Missile Defense," *Second Line of Defense* (April 20, 2018), https://sldinfo.com/2018/04/the-us-army-innovation-and-shaping-a-way-ahead-for-missile-defense/
4. https://www.3rdmaw.marines.mil/News/News-Article-Display/Article/1120033/wti-raises-the-bar-with-new-curriculum/

6. 2020 Interviews and Visit

1. Robbin Laird, *The Coming of the CH-53K: A New Capability for the Distributed Force* (2023), 232 pages.
2. Robbin Laird, *My Fifth Generation Journey: 2004-2018* (2024), 375 pages, chapter one.
3. See the argument developed in our book, *A Maritime Kill Web Force in the Making: Deterrence and Warfighting in the 21st Century* (2023), 310 pages.
4. Robbin Laird, "Colonel Jack Perrin on the CH-53K Program: An Update,"

Notes

Defense.info (February 25, 2020), https://defense.info/interview-of-the-week/colonel-jack-perrin-on-the-ch-53k-program-an-update-on-a-key-program/.
5. Robbin Laird, "Digital Interoperability and Kill Web Perspective for Platform Modernization: The Case of the Viper Attack Helicopter," *Second Line of Defense* (June 16, 2020), https://sldinfo.com/2020/06/digital-interoperability-and-kill-web-perspective-for-platform-modernization-the-case-of-the-viper-attack-helicopter/

7. 2023 Interviews and Visit

1. Robbin Laird, *Training for the High-End Fight: The Strategic Shift of the 2020's* (2020) and Robbin Laird and Edward Timperlake, *A Maritime Kill Web Force in the Making: Deterrence and Warfighting in the XXIst Century* (2022).
2. Megan Eckstein, "Marines Ditch MUX Ship-Based Drone to Pursue Large Land-Based UAS, Smaller Shipboard Vehicle," *USNI News* (March 10, 2020), https://news.usni.org/2020/03/10/marines-ditch-mux-ship-based-drone-to-pursue-large-land-based-uas-smaller-shipboard-vehicle
3. https://www.mca-marines.org/wp-content/uploads/0520-DASC.pdf.
4. https://www.marinecorpstimes.com/news/your-marine-corps/2020/10/12/marine-corps-creates-new-mos-for-mq-9-reaper-pilots/
5. https://www.marcorsyscom.marines.mil/News/News-Article-Display/Article/2735502/marine-corps-successfully-demonstrates-nmesis-during-lse-21/.
6. Robbin F, Laird, and Edward Timperlake. *A Maritime Kill Web Force in the Making: Deterrence and Warfighting in the 21st Century* (pp. 109-110). Kindle Edition.
7. https://sldinfo.com/2023/02/a-mission-kitable-aircraft-for-kill-web-operations-colonel-marvel-discusses-the-way-ahead-with-the-osprey/.
8. https://www.marines.mil/News/News-Display/Article/2708120/expeditionary-advanced-base-operations-eabo/

9. Retrospectives

1. https://www.history.navy.mil/content/dam/nhhc/research/histories/naval-aviation/Naval%20Aviation%20News/2000/2002/jan-feb/tactics.pdf.
2. https://www.usmcu.edu/Portals/218/LtGen%20William%20H_%20Fitch.pdf.

Made in the USA
Middletown, DE
03 September 2024